NUMERICAL MATHEMATICS AND SCIENTIFIC COMPUTATION

Series Editors

G. H. Golub   R. Jeltsch   W. A. Light
K. W. Morton   E. Süli

NUMERICAL MATHEMATICS AND SCIENTIFIC COMPUTATION

*P. Dierckx: *Curve and surface fittings with splines*
*H. Wilkinson: *The algebraic eigenvalue problem*
*I. Duff, A. Erisman, and J. Reid: *Direct methods for sparse matrices*
*M. J. Baines: *Moving finite elements*
*J. D. Pryce: *Numerical solution of Sturm–Liouville problems*
K. Burrage: *Parallel and sequential methods for ordinary differential equations*
Y. Censor and S. A. Zenios: *Parallel optimization: theory, algorithms and applications*
M. Ainsworth, J. Levesley, M. Marletta and W. Light: *Wavelets, multilevel methods and elliptic PDEs*
W. Freeden, T. Gervens, and M. Schreiner: *Constructive approximation on the sphere: with applications to geomathematics*
J. W. Jerome: *Modelling and computation for applications in mathematics, science, and engineering*

Monographs marked with an asterisk (*) appeared in the series 'Monographs in Numerical Analysis' which has been folded into, and is continued by, the current series.

Modelling and Computation for Applications in Mathematics, Science, and Engineering

Edited by

JOSEPH W. JEROME

Department of Mathematics
Northwestern University

CLARENDON PRESS · OXFORD
1998

Oxford University Press, Great Clarendon Street, Oxford OX2 6DP
Oxford New York
Athens Auckland Bangkok Bogota Bombay
Buenos Aires Calcutta Cape Town Dar es Salaam
Delhi Florence Hong Kong Istanbul Karachi
Kuala Lumpur Madras Madrid Melbourne
Mexico City Nairobi Paris Singapore
Taipei Tokyo Toronto Warsaw
and associated companies in
Berlin Ibadan

Oxford is a trade mark of Oxford University Press

Published in the United States by
Oxford University Press Inc., New York

© Oxford University Press, 1998
Chapter 4 © Todd F. Dupont and A. E. Hosoi

All rights reserved. No part of this publication may be reproduced,
stored in a retrieval system, or transmitted, in any form or by any means,
without the prior permission in writing of Oxford University Press.
Within the UK, exceptions are allowed in respect of any fair dealing for the
purpose of research or private study, or criticism or review, as permitted
under the Copyright, Designs and Patents Act, 1988, or in the case of
reprographic reproduction in accordance with the terms of licences
issued by the Copyright Licensing Agency. Enquiries concerning
reproduction outside those terms and in other countries should be
sent to the Rights Department, Oxford University Press,
at the address above.

This book is sold subject to the condition that it shall not, by way
of trade or otherwise, be lent, re-sold, hired out or otherwise circulated
without the publisher's prior consent in any form of binding or cover
other than that in which it is published and without a similar condition
including this condition being imposed on the subsequent purchaser.

A catalogue record for this book is available from the British Library

Library of Congress Cataloging in Publication Data
(Data available)

ISBN 0 19 850080 7

Typeset using LaTeX
Printed in Great Britain by
Biddles Ltd., Guildford & King's Lynn

Preface

This volume contains articles based on a workshop held at Northwestern University during the period 3–4 May 1996. They are organized into nine chapters. Given the theme of the volume, that of applications of modelling and computation to mathematics, science, and engineering, the first article presented is by a theoretical and computational chemist, Mark Ratner. Professor Ratner makes the convincing point that computational chemistry simulations are among the most computer-intensive programs currently employed. He speaks conversationally and factually to this and related issues of modern chemistry, particularly as regards the modelling of long-range molecular forces. An illustrative example, in terms of ion transport in polymer electrolyte systems, is presented. His article represents a challenge to mathematicians and chemists to work interactively in the modelling of molecular science. Deterministic as well as stochastic models are discussed.

Modelling biological and chemical contamination is the subject of the next chapter by Mary Wheeler, Clint Dawson, and Joe Eaton. Professors Wheeler, Dawson, and Eaton present sophisticated convection–reaction–diffusion systems in the modelling of subsurface flow and transport in heterogeneous porous media, to which are applied state-of-the-art computational procedures. A noteworthy aspect of this article is its inclusion of hierarchical ideas from computer science in a problem-solving environment, a few steps in the direction of expert systems.

The third article, by Yanmu Zhou, is a change of pace, in that it deals with discrete mathematics, the efficient placement of points on the sphere. Connections to the buckyball structure of carbon 60 are made by Professor Zhou, as well as to the fascinating area of fullerenes more generally. Discrete electrostatics and the packing problem also arise in this study. Algorithms for efficient placement are presented and displayed.

The fourth chapter, by Todd Dupont and A. N. Hosoi, deals with dimension reduction techniques in incompressible fluid mechanics. Professor Dupont and Dr Hosoi have investigated in detail the interface between stability and instability in various pattern formations, specifically as realized in Rayleigh–Bénard convection, colloidal sedimentation, and droplet break-off.

The next two articles deal with aspects of charge transport. In the fifth chapter, Irene Gamba and Cathleen Morawetz analyse models which conveniently bridge compressible fluids (gas dynamics) and semiconductors. Indeed, Professors Gamba and Morawetz have studied the natural generalization of two-dimensional, steady, irrotational, compressible, viscous flow in a channel, to the case when the fluid is charged. In the final section of this article is a discussion of various semiconductor models.

The sixth chapter, by Gui-Qiang Chen, the editor, Chi-Wang Shu, and Dehua Wang, deals with two-carrier charge transport, coupled in hydrodynamic system format. For example, the Gunn diode is naturally seen as the realization of two carrier species, differentiated by energy valleys. This represents a highly oscillatory system. Terms encompassing higher-dimensional symmetry are included in this hydrodynamic system. This allows for symmetry breaking in the MESFET transistor, for example. The Gunn oscillator and the MESFET are seen as prototypical applications for the modelling and analysis.

The last three articles deal with theoretical issues which have significant impact upon computing. In the seventh chapter, H. Thomas Banks and Richard Fabiano discuss the approximation problem in control theory. Professor Banks and Professor Fabiano have discovered that approximation schemes which converge nicely for the direct (simulation) problem do not necessarily converge for the parameter estimation or optimal control problems. This is a major idea in optimization. The associated investigation of the equivalence of consistency and stability with convergence is a central problem in the theory of numerical processes.

In the eighth article, I. E. Pritsker and Richard Varga study deep questions related to weighted approximation by polynomials in the complex plane. Professors Pritzker and Varga invoke surprising connections to potential theory in this discussion. Open questions involve extensions to rational approximation. Earlier work of the second author established connections to the heat equation attendant in such questions.

In the ninth chapter, Gui-Qiang Chen, the editor, and Bo Zhang establish the drift-diffusion limit, for hydrodynamic systems, as high-friction idealization. This is the first time this has been carried out for the full hydrodynamic model. The article includes a special case, worked out in detail, to motivate the general case. The article is a theoretical investigation of the idea, well known to physicists, that high scattering rates are associated with drift-diffusion models.

Evanston, Illinois J.W.J.

November 1997

Contents

List of contributors		xi
1	**Mathematics, Computational Chemistry, and Battery Building: Some Notions** *Mark Ratner*	
	1 Introduction	1
	2 Mathematical Challenges from Theoretical and Computational Chemistry	2
	3 Molecular Dynamics Algorithms and Long-Range Forces	5
	4 An Example: Polymer Electrolyte Simulations	8
	4.1 Physical Background: Polymer Electrolytes	9
	4.2 Molecular Dynamics Simulations in Concentrated Electrolyte Solutions	9
	4.3 Polymer Electrolytes: Dynamic Percolation Models	13
	5 Remarks	20
	Acknowledgements	21
	Bibliography	21
2	**Transport of Multispecies Contaminants with Biological and Chemical Kinetics in Porous Media** *Mary F. Wheeler, Clint Dawson, and Joe Eaton*	
	1 Introduction	25
	2 Model Formulation	26
	2.1 Black-Oil Model	29
	3 Mathematical and Algorithmic Issues	30
	3.1 Algorithms	31
	4 A Problem-Solving Environment for Flow in Permeable Media	33
	5 Numerical Results	34
	Acknowledgements	35
	Bibliography	35
3	**Equidistribution and Extremal Energy of N Points on the Sphere** *Y. M. Zhou*	
	1 Introduction	39
	2 Notation, Terminology, and Problems	40
	2.1 Various Problems of Arranging Points on the Sphere	41

3	Equal-Area Partitions and Extremal Energy Estimates	42
4	Numerical Experiments, Asymptotics of Extremal Energy	44
5	Geometry of Extremal Points and Carbon Fullerenes	45
	5.1 Carbon Fullerenes	45
	5.2 Geometry of the Extremal Points	47
6	A Class of Explicit Points that Equally Distribute over the Surface of the Sphere	48
	6.1 Construction of the Generalized Spiral Points	49
	6.2 How Good are the Generalized Spiral Points?	50
	6.3 A Variant of the Generalized Spiral Points	52
Acknowledgements		54
Bibliography		55

4 Some Reduced-Dimension Models Based on Numerical Methods
Todd F. Dupont and A. E. Hosoi

1	Introduction	59
2	Rayleigh–Bénard Convection	60
	2.1 Reduction with One Temperature Trial Function	64
	2.2 Reduction with Two Temperature Trial Functions	65
	2.3 Numerical Solution and Results	66
	2.4 Linear Stability of the Reduced System	68
	2.5 Error Bound for Galerkin Approximations	69
3	Colloidal Shocks	72
	3.1 Differential Model	72
	3.2 Galerkin Reduction	73
	3.3 Computational Results	74
	3.4 Alternative Equations	76
	3.5 Round Tubes	76
4	Drops and Jets	77
5	Other Resources	80
Acknowledgements		80
Bibliography		80

5 Viscous Approximation to Transonic Gas Dynamics: Flow Past Profiles and Charged-Particle Systems
I. M. Gamba and C. S. Morawetz

1	Introduction	81
	1.1 Transonic Flow Equations	82
2	Presentation of the Problem in the General Case	84
	2.1 Potential Flow and the Choice of the Viscous–Friction Force Term $\vec{F}$	85
	2.2 Further Conditions	88
	2.3 The Domain and Boundary Conditions	90
	2.4 Existence and Uniform Bounds	91

		2.5	Conclusions about Transonic Flow	95

	2.5	Conclusions about Transonic Flow	95
	2.6	About the Semiconductor Device Model	96
3	The Semiconductor		97
Acknowledgements		98
Bibliography		98

6 Two-Carrier Semiconductor Device Models with Geometric Structure and Symmetry Properties
G.-Q. Chen, J. W. Jerome, C.-W. Shu, and D. Wang

1	Introduction		103
	1.1	Description of the Gunn Oscillator	104
	1.2	Basic MESFET Description	105
	1.3	A Well-Posed Reduced Model	106
2	The Mathematical Framework		108
	2.1	Preliminaries	108
	2.2	The Shock Capturing Scheme	112
3	Spherically Symmetric Solutions and Nozzle Solutions		115
	3.1	Uniform Estimates	116
	3.2	H^{-1} Compactness of Entropy Measures	122
	3.3	Convergence and Existence	125
	3.4	Nozzle Solutions	128
4	The Simulation of the Gunn Diode		129
5	The Simulation of the MESFET: Symmetry and Symmetry-Breaking		133
Acknowledgements			138
Bibliography			139

7 Approximation Issues for Applications in Optimal Control and Parameter Estimation
H. T. Banks and R. H. Fabiano

1	Introduction	141
2	The Simulation Problem	143
3	The Optimal Control Problem	144
4	The Parameter Estimation Problem	159
Acknowledgements		163
Bibliography		163

8 Zero Distribution, the Szegő Curve, and Weighted Polynomial Approximation in the Complex Plane
I. E. Pritsker and R. S. Varga

1	Introduction		167
2	Weighted Polynomial Approximation by $\{e^{-nz}P_n(z)\}_{n\in\mathbb{N}_0}$		168
3	General Weighted Polynomial Approximation		173
4	Applications		175
	4.1	Incomplete Polynomials and Laurent Polynomials	176

		4.2	Jacobi and Jacobi-Type Weights	177

	4.3	Exponential Weights	179
5	Some Proofs		179
6	Further Remarks and Open Problems		187
Bibliography			187

9 Existence and the Singular Relaxation Limit for the Inviscid Hydrodynamic Energy Model

Gui-Qiang Chen, Joseph W. Jerome, and Bo Zhang

1	Introduction		189
	1.1	Idealized Drift-Diffusion	190
	1.2	The Hydrodynamic Model	191
	1.3	Regimes Defined by Damping	193
2	Energy Method for a Scalar Equation with Damping		194
3	Global Smooth Solutions of the Inviscid Hydrodynamic Energy Model		199
	3.1	Local Solutions	201
	3.2	A Priori Estimates via the Energy Method	205
	3.3	Global Existence, Uniqueness, and Asymptotic Decay	210
4	Singular Relaxation Limit to Drift-Diffusion Equations		210
Acknowledgements			213
Bibliography			213

Contributors

H. Thomas Banks North Carolina State University, Raleigh, NC 27695.
Gui-Qiang Chen Northwestern University, Evanston, IL 60208.
Clint Dawson University of Texas, Austin, TX 78712.
Todd Dupont University of Chicago, Chicago, IL 60637.
Joe Eaton University of Texas, Austin, TX 78712.
Richard Fabiano University of North Carolina, Greensboro, NC 27412.
Irene Gamba New York University, New York, NY 10012.
A. N. Hosoi University of Chicago, Chicago, IL 60637.
Joseph Jerome Northwestern University, Evanston, IL 60208.
Cathleen Morawetz New York University, New York, NY 10012.
I.E. Pritsker Kent State University, Kent, OH 44242.
Mark Ratner Northwestern University, Evanston, IL 60208.
Chi-Wang Shu Brown University, Providence, RI 02912.
Richard Varga Kent State University, Kent, OH 44242.
Dehua Wang University of California, Santa Barbara, CA 93106.
Mary Wheeler University of Texas, Austin, TX 78712.
Bo Zhang Stanford University, Stanford, CA 94305.
Yanmu Zhou Northwestern University, Evanston, IL 60208.

1
Mathematics, Computational Chemistry, and Battery Building: Some Notions

Mark Ratner
Northwestern University

1 Introduction

Mathematics is the language of science. As soon as one moves beyond taxonomy and nomenclature, nearly all scientific disciplines (and, increasingly, those in the social sciences) aim to express their results in quantitative, predictable, theoretically understandable terms. To do so, they require the language, conceptual basis, and predictive power of mathematics. Although mathematics is the language of science, often the translation problem is vexing: for example, to formulate a correct theory of mechanics, Newton actually had to invent the appropriate mathematical language. Similarly, although the phenomenon of Brownian motion is common to everyone who has seen the dust in a sunbeam, its proper mathematical formulation came about only in the early 20th century (Einstein, Langevin, Perrin, and others) [41].

Often, these problems in using the correct mathematical language to describe physical phenomena arise because of lack of understanding of the physical phenomena. Less commonly, they occur because the appropriate mathematics is not yet formulated. Most commonly, they occur because the scientists involved simply don't know enough mathematics.

These issues have become exacerbated with the advent of that sleek, powerful, rapidly growing new weapon in the arsenal, computational science. The computer is really a transforming presence in the sciences, promising much, accomplishing a great deal, but also making many demands (for computational results to be meaningful, the problem must be well posed and the simulation techniques appropriate and correct) and offering lures, wiles, and distractions (there are many more incorrect than correct simulations).

The National Academy, in 1994, published a report on mathematics and computational chemistry [29]. While this report, like most others of its type, is to some extent incomplete, episodic and tendentious, it does point out some of the benefits that may accrue if computational scientists would learn more mathematics, and if mathematicians would interest themselves in the problems associated with computational science. The readers of this book probably know this already, but in this short article I would like both to stress the results of the

National Academy study and to provide one simple illustration of how important it is that appropriate mathematical understanding be brought to computational modeling. The adjectives used above to describe the National Academy report apply, with even greater strength, to this paper. Nevertheless, I hope that the clear benefits that some reasonable mathematical analysis has brought to the problem of long-range force calculation will serve as an illustration of the importance, and the impact, of appropriate mathematical understanding brought to the analysis of computational scientific problems.

Section 2 recalls some of the conclusions of the National Academy study, and outlines some of the areas in which important problems remain to be addressed. Section 3 focuses on the particular issue of long-range forces, and their proper treatment in simulations. It points out that the correct mathematical understanding goes back to the work of Ewald at the beginning of the century [16], that the problem was clearly formulated, and essentially solved, by applied mathematicians in the 1980s [12], but that the published literature as late as 1995 (and almost certainly as late as 1997!) is still making fundamental errors in the computational study of systems containing long-range forces [38, 33, 43, 39, 7, 8]. Section 4 discusses a particular physical situation, that of polymer electrolytes. This section begins with an introduction concerning the physical properties of such electrolytes, and then discusses modeling studies in which different types of computational study have been used to understand the mechanistic details of ion transport in polymer electrolyte systems.

2 Mathematical Challenges from Theoretical and Computational Chemistry

The title of this section is also the title of a report [29], published by the National Research Council in 1995, that was the product of a Committee on Mathematical Challenges from Computational Chemistry. This Committee (the list of members is incorporated as Fig. 1) consisted mostly of chemists, with some mathematicians. This over-weighting of chemists may partly be because of the selection process, but it also indicates the desire of the chemical community to increase the mathematical sophistication, appropriateness, and adequacy of their research. This report (available from the National Academy Press), focuses on several topics. The chapter headings are: the Emergence of Computational Chemistry, Examples of Constructive Cross Fertilization Between the Mathematical Sciences and Chemistry, Mathematical Research Opportunities in Theoretical/Computational Chemistry, Cultural Issues and Barriers to Interdisciplinary Work, and Conclusions and Recommendations. The most important data from the report are contained in the Executive Summary, and in Figs. 2 and 3.

The Executive Summary finishes with Conclusions and Recommendations. These include:

(1) Several notable success stories can be identified, illustrating the value of

Committee on Mathematical Challenges From Computational Chemistry

Frank H. Stillinger, AT&T Bell Laboratories, Chair
Hans C. Andersen, Stanford University
Louis Auslander, City University of New York
David L. Beveridge, Wesleyan University
Ernest R. Davidson, Indiana University
Wayne C. Guida, Ciba-Geigy Corporation
Peter A. Kollman, University of California at San Francisco
William A. Lester, Jr., University of California at Berkeley
Yvonne C. Martin, Abbott Laboratories
George C. Schatz, Northwestern University
Tamar Schlick, New York University and Howard Hughes Medical Institute
L. Ridgway Scott, University of Houston
DeWitt L. Sumners, Florida State University
Peter G. Wolynes, University of Illinois at Urbana-Champaign

Board on Chemical Sciences and Technology Liaison
Kendall N. Houk, University of California at Los Angeles

Scott T. Weidman, Study Director
Tana L. Spencerk, Project Assistant

FIG. 1. The membership of the Committee on Mathematical Challenges from Computational Chemistry, authors of ref. [29]

Time Used (%)	Application	Description
7.1	ESP	Molecular dynamics
6.7	Gaussian	Quantum chemistry
5.4	AMBER	Molecular dynamics
2.6	TREESPH	Galactic dynamics
2.1	GAMESS	Quantum chemistry
2.0	ARGOS	Molecular dynamics
1.5	CGCM	Coupled ocean–atmosphere global climate model
1.5	DMOL	Quantum chemistry
1.3	COULMETL	Materials science
1.2	DIEL	Materials science

FIG. 2. The top ten applications in terms of percentage of CRAY C90 usage at the San Diego Super Computer Center between December of 1993 and August of 1994. Note the clear dominance of computational chemistry. (From ref. [29])

	Quantum electronic structure	Molecular mechanics	Condensed-phase simulations	Density functionals	N-representability	Design of molecules	Construction of potential energy functions	Gas-phase dynamics	Polymers	Topography of potential energy surfaces	Biological macromolecules (including protein folding)
Adaptive and multiscale methods	H	M	M				M	M	M	M	
Special bases	H		M	M				H			
Differential geometry		M				M	M		M	H	H
Functional analysis	H		M	H	H		H	M		M	M
Graph theory	M	M	M			H			M	M	M
Group theory	H		M		M	M		M	M		M
Optimization	H	H	H	H	M	H	H	M	H	H	H
Numerical linear algebra	H	H		M			M			M	
Number theory		M						M		M	
Pattern recognition	M	M	H			H		M	H	H	H
Probability and statistics	H		H		M	H		M	H	H	H
Several complex variables	H		H	M				M		M	M
Topology	M	M	H			H	M	M	H	H	H
Dynamical systems		M	H					H	M	H	M

FIG. 3. Subjective assessment of potential cross fertilization between major areas of the mathematical sciences and theoretical and computational chemistry. H & M represent, respectively, high overlap and possible synergy. (From ref. [29])

interdisciplinary stimulation and synergistic research collaboration.
(2) Many opportunities exist for further collaborations between the mathematical and chemical sciences. The productivity of applied computational chemistry would likely be enhanced, which would be potentially significant for industry.
(3) Active encouragement and further collaboration are warranted.
(4) The cultural differences between the mathematics and chemistry communities have tended to act as barriers to collaboration.
(5) The overall volume in specialized technical literature aggravates the communication problems between fields and occasionally leads to wasted effort, redundancy, and rediscovery.

Actually, the Executive Summary and the Recommendations are more extensive, but I have selected this particular list because it speaks so directly to the concerns of this book: the physical sciences do indeed offer important challenges for mathematics. In many cases the mathematical analysis in back of extensive efforts in physical science research is inappropriate or inadequate, or both, and these fields offer interesting and unusual challenges for mathematicians. Nobody knows this better than the organizer of the workshop upon which this volume is based; Jerome's work in modeling semiconductor circuits [23], ion channel flow [10], and other response phenomena has been remarkable for its productivity, mathematical sophistication, and physical appropriateness. It is just that we need more of this sort of work. Figure 3, reproduced from the National Academy Report, suggests areas in which there are promising overlaps between theoretical chemistry and mathematics. The data is to some extent impressionistic, but the chemical topics along the top row are (in general) of major importance. For example, Fig. 2 shows the usage of supercomputer time in 1994, at the San Diego Super Computer Center. Notice that of the eight most used programs, six involve theoretical chemistry. The topics of quantum chemistry, molecular dynamics, and molecular modeling are areas in which appropriate mathematics is necessary, but too often absent. Without reproducing the entire Table of Contents of the National Academy Report, we note that there are several topics that are discussed at some length. These include molecular dynamics algorithms—that is, algorithms for numerically integrating the Newton equations of motion, subject to appropriate boundary conditions and potentials. Sub-topics include enhanced sampling, numerical methods, symplectic integrators, time scale problems, implicit integration schemes. There is also an extended section on multivariant minimization, including issues such as problem classification, complexity, global optimization, and saddle point location.

I mention these topics simply to give a flavor for why this area is important to theoretical chemistry, and of the extensive national demand to understand behaviors. Consider, for example, the multi-particle potential energy surface corresponding to, say, a glass-forming material such as a protein. The classification and understanding of the multiple minima problem so defined are very important for theoretical chemistry, and very important for life: this is supposed to represent the potential energy surface of a protein, and the native conformation corresponds to some subset of the multiple minima surface (generally in roughly 10^5 dimensions) that, for example, the rhodopsin in the eyes reading this paper contain. Understanding the nature of the minima, the dynamical sampling of the minima, the averaged configurations, and their properties is one of the great problems in theoretical chemistry and biology today.

It is true that, for the mathematicians reading this article, the labels on the left in Fig. 3 will be most transparent, while for the chemists it will be the columns along the top. It is in the interaction between the columns and rows that, to some great extent, real progress is to be made.

3 Molecular Dynamics Algorithms and Long-Range Forces

One striking result of the availability of computational resources in the chemical community has been the ballooning of simulation studies. The most common simulation method for chemists, and in many ways the most elegant and useful, is the technique now referred to as molecular dynamics [21]. In a molecular dynamics algorithm, one simply solves the Newton equations,

$$m_i \ddot{\vec{x}}_i = -\nabla \mathbf{V}_{\text{TOT}}(\vec{x}), \qquad (3.1)$$

for a number of interacting particles. Once the size of the simulation (the number of particles of each sort), the total potential $\mathbf{V}_{\text{TOT}}$, and the appropriate boundary conditions are defined, and an appropriate integration method is selected, the molecular dynamics (MD) simulation should proceed smoothly. The technique was first introduced in the 1950s and 1960s. Not surprisingly, the first studies were from national laboratories (Berkeley [2], Argonne [34], and Brookhaven [48]): they were the ones with the computational resources. The first studies were on systems with short-range potentials, particularly hard sphere and hard disk potentials (in California) and soft, short-range potentials (at Argonne). These early studies were remarkable in their intuitiveness, importance, intelligence, and influence.

For pair potentials that are short-ranged (that is, potentials that fall off as $r^{-s}, s > 3.0$), simulations are relatively straightforward. In particular, the short range of the potential means that the force calculation, that drives the time evolution of the algorithm, can be completed efficiently, and that the force is convergent as a function of system size.

For long-range forces, however ($r^{-s}, s < 3.0$), the potentials are conditionally convergent. To take the physically most important case, in a situation with charged particles, the coulomb interaction is

$$V(\vec{r}_i, \vec{r}_j) = \frac{q_i q_j}{|\vec{r}_i - \vec{r}_j|}, \qquad (3.2)$$

and the overall coulomb potential felt by an individual particle is simply

$$v_i(\vec{r}_i) = \sum_{j \neq i} V(\vec{r}_i, \vec{r}_j). \qquad (3.3)$$

There are two components to this equation: the interaction with the particles of like charge and that with particles of unlike charge. In the physical system, the net charge vanishes, so that the total interaction potential is finite in the thermodynamic limit. The component parts, however, are each formally divergent: that is, as the number of charged particles increases with the volume of the system, the overall potential is a conditionally convergent sum, with the three physical components (the interaction of all plus-charged particles with one another, the interaction of all negative-charged particles with one another, and the plus–minus interaction) each divergent.

The proper analysis of this problem was originally introduced by Ewald [16]. Its extension to the evaluation of the force in molecular dynamics simulations was explained [12], with clarity, mathematical rigor, and completeness, by deLeeuw, Perram and Smith in 1980. Essentially, one reorganizes the sum in such a way that the computation is both efficient and convergent. To do so, one separates the interactions into a part treated in direct space (for short-range interactions) and a part treated in Fourier space (for long-range interactions). The space cut-off distance between the two is a convergence parameter of the calculation.

The important point is that, if one uses appropriate (periodic) boundary conditions and calculates the interactions using the Ewald sum, a number of pathologies that would otherwise affect MD simulation are avoided [12, 38, 33, 43, 39, 7, 8, 21]. The computational effort is a bit greater than simple potential truncation, but it gives the right answer!

A number of simulations followed the original paper by deLeeuw, Perram, and Smith, and it became quite clear that the Ewald approach really does permit simulations in coulomb systems.

This point, however, did not percolate very well or very effectively within the simulations community. References [38, 33, 43, 39, 7, 8] are examples of recent simulation literature in which it has been rediscovered that long-range potentials give problems, that truncation of the long-range potential in an arbitrary fashion gives a discontinuity in the force and can lead to artifactual results, and that these effects are important. The error involved in the truncation is a slightly insidious one: it seems clear from "physical intuition" that if one doubles the size of the simulation box, and finds essentially the same result, then the simulation is independent of the box size, and one can relax. This is incorrect: as equation (3.4) shows, the divergence in the self-interaction of the like-charge components will not appear for any finite box - this pathology is not expected on physical grounds, and really arises from the very long-range of the potential.

$$\int_0^\infty \frac{1}{r} d(volume) = \int_0^\infty \frac{4\pi}{r} r^2 \, dr = \int_0^R \frac{4\pi}{r} r^2 \, dr + \int_R^\infty \frac{4\pi}{r} r^2 \, dr \qquad (3.4)$$

We see that the divergence of the integral occurs only in the limiting sense, i.e., only in the second term in (3.4).

Does it matter? Yes. For example, a well-cited and important simulation study by a group at Berkeley [4] on the interaction of positively charged iron ions in aqueous solution used truncated potentials, without proper analysis or treatment of the long-range part of the potential. The results that they obtained, which were published in the most prestigious journal in the field, very broadly cited and used by a number of experimentalists in interpreting their results, are largely artifactual: in particular, they observe the (unphysical) result that these two positively charged iron ions form a complex with one another. This result was unanticipated on physical grounds, and made a bit of a splash at the time. Re-examination of this problem, by use [5] of appropriate Ewald boundary conditions, shows that this complex simply doesn't exist: the iron ions really do effectively repel, and do not form a bound complex.

Indeed, this error had an even more deleterious effect on the field, because the result of approximate statistical mechanical closure theories gave [48] the same artifactual result as the incorrect simulation! This has now been worked out quite carefully for the ions in solution problem, but in a number of other simulation systems (such as polyelectrolyte binding, cluster formation or protein interactions), this incorrect treatment of long-range potentials persists, and causes trouble. This point is detailed, occasionally with pain, in references [38, 33, 43, 39, 7, 8, 21].

This story of the appropriate treatment of long-range potentials in molecular dynamics systems serves nicely to illustrate the importance of appropriate mathematical analysis in theoretical chemistry, the mathematical challenge involved (the original paper [12] by deLeeuw, Perram, and Smith runs to many pages of the Proceedings of the Royal Society), the importance of establishing the correctness of the approaches used both in theory and in simulation, and the difficulty of communications across the mathematics–chemistry border.

Many other comparable examples can be found in the literature. Occasionally, things work to spite some mathematical concerns (for example, the Stirling approximation is often used by physical scientists in situations in which the argument of the gamma function is not large enough for the approach to be valid, stationary-phase approximations are often employed in reduced-dimensionality situations for which the stationary-phase conditions probably do not apply [40], limiting processes are often taken with complete abandon, etc.). Physical scientists will often indulge in such behavior because "physical intuition suggests that this is acceptable, although the mathematics is not clear;" the Ewald example above shows how dangerous this can in fact be.

4 An Example: Polymer Electrolyte Simulations

In all battery systems, current is carried between two electrodes by ionic motion through an electrolyte. One particularly interesting class of systems involves so-called polymer electrolytes—that is, electrolytes in which ions move in a host medium consisting of a polar polymeric material [17, 26, 19, 18, 36, 46, 25, 3, 45]. These systems are quite complicated: they are dense materials, so that one expects a Stokes–Einstein-like relationship between diffusion and viscosity [42]. The relaxation processes of the polymer host occur over many time scales [17, 26, 19, 18, 36, 46, 25, 3], The structure is non-crystalline, the potentials of interaction are both short-range (within the polymer) and long-range (among the ions). Because these materials are of substantive interest for producing better batteries (for a number of reasons, including low cost, long shelf life, high energy density, dimensional capability, light weight, chemical tailorability) [17, 26], understanding the nature of charge transport within the polymer electrolyte and how it can be varied with chemical substitution and appropriate laboratory parameters is an important endeavor.

Two of the many approaches taken to understanding ion transport in polymer electrolytes are based on molecular dynamics for short-time behavior [32, 28, 30], and on a generalized hopping model involving mixed (or stirred, or dynamic) percolation theory for long-time behavior [13, 15, 14, 20, 35].

4.1 Physical Background: Polymer Electrolytes

The simplest polymer electrolytes are made by dissolving a uni/univalent salt, such as NaI, in a polar polymer. A typical polar polymer is polyethyleneoxide, whose chemical formula is shown in Fig. 4. Polyethyleneoxide will dissolve a

$$(-CH_2 - CH_2 - O-)_n \quad PEO$$

FIG. 4. Polyethyleneoxide. Complexes between this polymeric species and simple uni–univalent salts such as LiI produce conductive polymer–salt complexes.

number of polar salt guests, and the resulting complexes exhibit ionic conductivity that is more than 10 000-fold increased over the (extrinsic) parent polymer. These complexes are of interest as electrolyte materials.

For these complexes, the concentration of the ions tends to be quite high (roughly 0.1–1.0 molar), so that the average distance between ions is of the order of four or five times the ionic radius, and the local effective coulomb interaction between ions is substantially greater than thermal energy. Under these conditions, the standard theories for dilute ionic solutions [42, 9, 1, 31] fail badly, and understanding either the structure or the transport becomes a complicated problem, characterized by long-range potentials in a dynamically disordered medium.

Of the many questions that might be asked about such a system, I would like to concentrate here on three: these are (1) Does ion pairing or clustering increase or decrease as the temperature is varied? (2) How does the relaxation process in the polymer host material modulate the conductivity, or diffusion coefficient, of the ions? (3) What suggestions for synthesis of better conductive materials can be gleaned from appropriate simulation studies?

We will briefly describe simulation studies that aim to answer, at least in part, these questions.

4.2 Molecular Dynamics Simulations in Concentrated Electrolyte Solutions

To deal with concentrated electrolytes, one can use MD simulation [32, 28, 30]. Clearly, the electrolyte nature (positive and negative ions in a polymer solution) will require the Ewald analysis (or another method for dealing properly with long-range forces) if the calculations are to be convergent. Molecular dynamics itself is most useful in the calculations that are completed at equilibrium, because one can then use analysis of correlation functions to examine transport coefficients, and calculations of position to examine structure [21]. With polymeric systems, obtaining the equilibrium state is extremely complicated: polymer relaxation times extend to the minutes time scale, and since the characteristic time scale of the molecular dynamics simulation is fixed by the curvature of the potential, appropriate time scales in most molecular systems are of the order of 10^{-15} seconds. It is therefore out of the question to simulate a polymer system long enough to attain true equilibrium. Since, however, many of questions the issue in polymer electrolytes pertain to any strong, concentrated electrolyte system, one can use simulations of simpler systems to gain some insight into the questions that arise in those materials. We therefore completed MD [32] simulations on solutions of NaI in simple ethers, rather than the polyether (PEO) of Fig. 4.

The details of the simulation are presented elsewhere [32]. The important point is that one uses the Ewald construction [12], with appropriately chosen interaction potentials, box sizes, integration algorithms, sampling methods, convergence tests, and human resources.

To answer the question of the temperature dependence of ion pairing or clustering, one can calculate the potential of mean force acting between two ions. The potential of mean force is essentially the single-body potential that reproduces the effective two-body distribution function, calculated from equilibrium simulations. Formally, it is related to the free energy, A, of ion interaction by the equation [41, 32, 24],

$$U - TS = A, \qquad (4.1)$$
$$A = -RT \ln g(r) = w(r). \qquad (4.2)$$

Here U, T, S, A, R, and w are respectively the energy, temperature, entropy, free energy, gas constant, and potential of mean force. The distribution function $g(r)$ is found directly from simulation. By running MD simulations at different temperatures, one can determine both of the components of this equation, and can study both the actual potential of mean force (that is, whether the ions have strong enough attraction to cluster or not) and its components in terms of entropic and enthalpic behaviors.

Experimentally, there are divergent claims. Work using magnetic resonance spectroscopy and light scattering suggests that at higher temperatures clustering occurs [49], and the salt actually precipitates out from the polymer electrolyte host. Simple considerations based on binding potentials, on the other hand [22], suggest that an ion pair will break up, so that clustering should be reduced with increasing temperature. Both of these suggestions appeared extensively in the experimental literature before the simulations discussed here were completed.

The simulations [32], and later experiments, are quite definite: in these materials, clustering increases with increased temperature. One can see this from snapshots of the structure (Fig. 5), or more clearly from calculations of the potentials of mean force (Fig. 6). Essentially, both the entropy and the energy terms give ion clustering, and with increased temperature the effective entropy is actually higher for the segregated materials than for the dissolved salt. This is due to the configuration entropy of the neat solution: ions will order any protic solvent (in this case an ether, but the same effect would be observed with water). Earlier simulations [32], in which the solvent was represented in terms of a point dipole in a spherical soft atom, did not observe increase of clustering with temperature: this is because the configurational entropy contribution is missing, since atoms have no entropy associated with rotations or orientation.

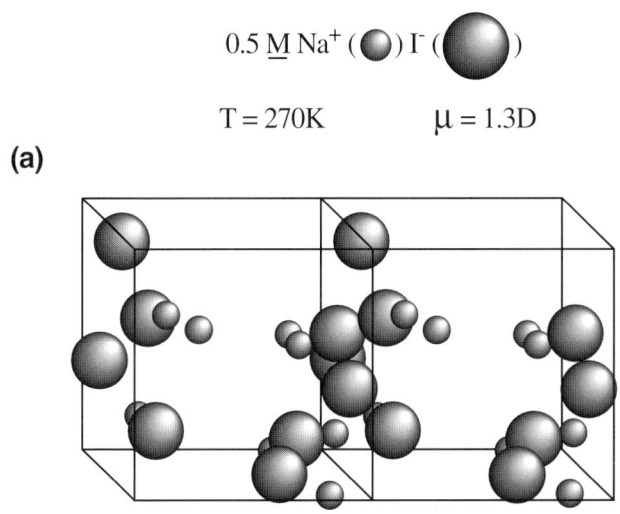

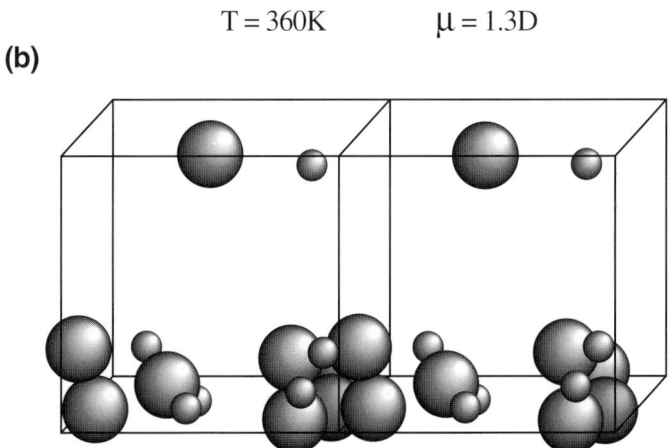

FIG. 5. Snapshots of a molecular dynamics simulation of the polymer electrolyte formed between sodium iodide, NaI, and a dielectric particle solvent. In Fig. 5a, the cations (small balls) and anions (large balls) are disordered, whereas they are clearly clustered in Fig. 5b. The simulations were completed at 270 K and 360 K, respectively: the results strongly suggest that ion pairing and clustering increase with increase in temperature. Solvent is omitted for clarity. (From ref. [32])

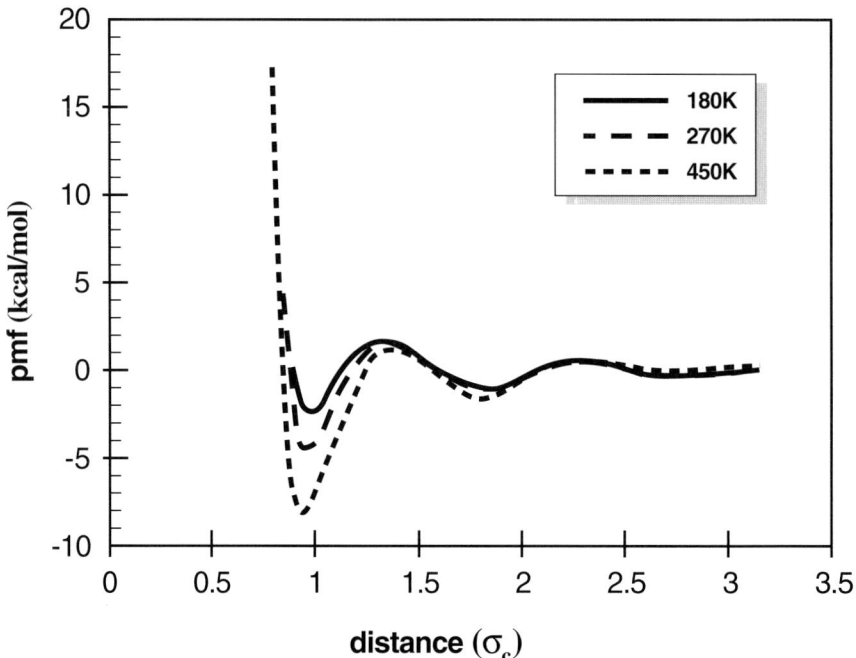

FIG. 6. The calculated potential of mean force acting between Na and I, obtained from molecular dynamics simulations. Notice the increasing depth of the first minimum, corresponding to the nearest neighbor pairs, with increase in temperature. This suggests, as does Fig. 5, that at increased temperature, clustering increases. (From ref. [32])

This result of simulations has been important in our experimental understanding of polymer electrolyte systems. Still, these results are really not for polymer electrolytes. There have been some molecular dynamics simulations on oligomeric representatives of polymer electrolytes, but the problem is a complicated one [28, 30]. Some simple dimensional analysis suggests the magnitude of the problem: if the molecular dynamics time step is 10^{-15}/sec, and one is dealing with a typical polymer electrolyte whose diffusion coefficient is 10^{-6} cm^2/s, with a characteristic diffusion coefficient of 10^{-7} cm^2/s, diffusive behavior suggests that the time required to observe the particle diffusing 5 Å (that is, roughly 3 times its own radius) is

$$\tau \approx \ell^2/D = 25 \cdot 10^{-16}\,\text{cm}^2/10^{-6}\text{cm}^2/\text{s} = 25 \cdot 10^{-10}\,\text{s}. \qquad (4.3)$$
$$\text{number of time steps} \approx 25 \cdot 10^{-10}\,\text{s}/10^{-15}\,\text{s} = 2.5 \cdot 10^6\,\text{steps} \qquad (4.4)$$

This means that one has to integrate the system for roughly 10^6 time steps to observe one diffusive motion! For statistically significant results to be obtained from this simulation, this is not possible using straightforward molecular dynamics.

The molecular dynamics method has been used [32, 28, 30] successfully to learn some things about polymer electrolytes: the results given here for ion clustering and pairing are an example. Other examples include the nature of the coordination site around the cation, the vibrational frequency distribution on the backbone lattice, and some information about the time scales appropriate for diffuse hops.

There are many difficulties entailed in doing any molecular dynamics simulation; the most common of these is the potential itself, which is never known with complete accuracy. Many incorrect results in the literature have been obtained by using inadequate potentials, but the general rule of thumb seems to be that if a behavior depends on the fine details of the potential, that behavior is probably not believable. From the point of view of mathematics in chemistry, however, the important thing is that issues like time stability, correct treatment of boundary conditions, sensitivity to rate of growth of chaotic behavior, etc., all depend on the details of the MD simulation; some mathematical analysis has been done here [12], but a great deal more is necessary to make molecular dynamics the robust, essentially exact treatment that it should become.

4.3 Polymer Electrolytes: Dynamic Percolation Models

Because of the time scale limitations just discussed, molecular dynamics does not provide an attractive interpretative language for understanding conduction in polymer electrolytes. Questions of interest to most chemists in studying such systems [17, 26, 19, 18, 36, 46, 25, 3, 45], including the mechanism of transport, the effects of pressure, differing choice of ions, differing polymer lengths, differing chemical substituent groups on the polymer, different temperatures, and chemical additives, simply cannot be approached using molecular dynamics because of extensive computational demand. Moreover, some mechanistic aspects are probably better appreciated using a simulation scheme that relates more directly to what is of interest (the ion motion) and less to the detailed motions of all the particles comprising the system (as would be derived from molecular dynamics simulation).

A number of experimental studies [17, 26, 19, 18, 36, 46] suggest that there is a Stokes–Einstein relationship [42] between the local microviscosity, η_{mic}, and the ion diffusion coefficient. This can be expressed as

$$\eta_{\text{mic}} \times D = \text{constant}. \tag{4.5}$$

The viscosity for glass-forming materials has been fitted extensively [17, 26, 19, 18, 36, 46] to the purely empirical so-called WLF relaxation law,

$$1/\eta_{\text{mic}} = C \, \exp(-B/(T - T_0)). \tag{4.6}$$

Here C, B, T, and T_0 are respectively a constant (the high-T limiting value of the inverse viscosity), an activation energy-like term, the Kelvin temperature, and the so-called equilibrium glass transition temperature. This last is the temperature at which the polymer becomes glassy under extremely slow cooling conditions. The Stokes–Einstein relation suggests that a model can be built in which ions hop diffusively on a lattice, but that the hopping rate depends on the local relaxation properties of the polymeric material; those relaxation properties themselves should be given by a variant of the WLF equation.

Such a model, called the dynamic bond percolation model, has been developed and extensively applied in the study of polymer electrolytes [17, 13, 15, 14, 20, 35]. We mention here some of the interesting mathematical aspects of this behavior. One can define a probability $P_j(t)$ to occupy lattice site j at time t, and one can define hopping rates $W_{j \to i}(t)$ as the local hopping rate from site j to site i at time t. One can then write a set of coupled differential equations, the so-called master equations, as

$$\dot{P}_j(t) = \sum_i \{W_{i \to j} P_i(t) - W_{j \to i} P_j(t)\}. \tag{4.7}$$

The hopping probabilities W are chosen from a distribution of hopping probabilities. In the simplest model, one assumes a cubic lattice, limits the hopping to nearest neighbors, and chooses a binary distribution such that

$$W_{i \to j} = \begin{cases} w & \text{probability } f \\ 0 & \text{probability } 1 - f. \end{cases} \tag{4.8}$$

Here f is the probability of a jump being allowed between two sites. In static percolation theory [44], the W's are independent of time: the lattice is assumed to consist of a series of jumps, some of which are permitted (rate w) and others are forbidden ($W = 0$).

Polymer electrolytes are not well discussed using static percolation theory, because the lattice is not static. Rather, the ions move in a dynamically disordered medium: that is, above the glass transition temperature, the viscosity follows the WLF rule of Eq. (4.6), so that the local environment in which the ion finds itself evolves on a characteristic time scale. This characteristic time scale, determined by the polymer solvent relaxation, is called the characteristic relaxation time, or (in the context of dynamic percolation models) [13, 15, 14, 20, 35] the renewal time, τ_{ren}. The physical picture of dynamic percolation, then, is that the bond probability assignments $W_{i \to j}(t)$ evolve in time, with the characteristic evolution time τ_{ren}.

Within this general dynamic percolation model, a number of different scenarios can be investigated: the renewal times can be correlated or uncorrelated, they can be chosen from a distribution or simply assigned a single value, they can be correlated with the ion position or independent of the ion position, and their distribution can be chosen in various ways (δ process, Poisson process, and various choices for the waiting time distribution) [14, 20].

The evolution of the ions following the master Eq. (4.7), with the percolation choice of Eq. (4.8) for the probabilities, but with probabilities reassigned on characteristic time scale τ_{ren}, has been investigated, and applied to the problem of polymer electrolyte diffusion [17, 13, 15, 14, 20, 35]. Because this model is an heuristic one, based on a series of physical and simplifying assumptions, it lacks the directness of the molecular dynamics simulation. On the other hand, it studies directly a physically interesting property, the ion motion.

The mathematics involved in the renewing, or stirred, percolation problem actually has some quite nice aspects of its own. This includes the so-called renewal time paradox [47], and the relationship to continuous time random walks [27]. Our purposes here are both served by briefly outlining some of the results that follow from straightforward analysis of the dynamic bond percolation problem as discussed above. Several important formal results have been obtained. These include [13, 15, 14, 20, 35]:

(1) Although a percolation threshold exists in static percolation models (if the probability f of Eq. (4.8) is smaller than a particular threshold value, f_{th}, then no diffusion is seen), in dynamic percolation theory the system is *always* diffusive: that is, the mean square displacement on a lattice of integer dimensionality for a dynamic percolation model is always linear in time. This is illustrated from some simulations, using a two-dimensional square lattice and a uniform distribution of renewal time intervals, in Fig. 7.

(2) There exists an analytical continuation formula, which effectively relates the frequency-dependent diffusion coefficient in the renewing case to the frequency-dependent diffusion coefficient, D_0, in the static percolation problem via

$$D(\omega) = D_0(\omega - i/\tau_{\text{ren}}). \qquad (4.9)$$

(3) In a simple one-dimensional model, one can actually derive the closed-form expression for the diffusion coefficient, given as

$$2D = \overline{<x^2>}/\tau_{\text{ren}}. \qquad (4.10)$$

Here the numerator is the averaged mean square displacement of the particle for a fixed renewal interval, and the bar implies average over the distribution of renewals. One suspects that similar behavior should hold in high dimensionality, but there is no proof.

(4) The model produces the appropriate limiting cases: when the renewal time becomes infinite, the results of static percolation theory follow directly. When the renewal is very rapid, the diffusion is simply that on a continuous lattice with a lower percentage, f, of available bonds.

(5) The restriction that the particles lie on a lattice can be relaxed, without significant effect on the predicted diffusion behavior.

(6) The process of renewal can either increase or decrease the overall diffusion coefficient. In the case (relevant for polymer electrolytes) in which motion within the renewal interval is limited by the availability of new sites, so that the plot of mean square displacement within a renewal interval is sublinear

in time, the renewal process increases the diffusion coefficient compared to the non-renewing situation (illustrated in Fig. 8). Conversely, if the mean square displacement is superlinear within each renewal interval, as will hold for ballistic transport for quantal electronic conduction in a good metal, the renewal process will reduce the overall diffusion coefficient—this is precisely what occurs in good metals. These behaviors are illustrated in Fig. 8.

(7) These results are largely independent of the details of renewal time distribution, although some pathologies can occur and, in particular, strong correlations between ion position and the renewal process can result [20] in substantially varied diffusion behavior.

Based on some formal analysis of the problem, and taking into account the lesson that long-range forces must be treated appropriately, extensive work has been published on modeling dynamical hopping in polymer electrolyte systems [13, 15, 14, 20, 35]. The most important result of these simulations is shown in Fig. 9, where the two lines represent the calculated diffusion coefficient as a function of the average renewal time. As Eq. (4.10) suggests (although the current calculations are for three-dimensional cubic lattices, not for a linear chain), the diffusion coefficient scales roughly with the inverse average renewal time: this is in agreement with the Stokes–Einstein behavior of Eq. (4.5). The other important result is that ion concentration or correlation effects will reduce the effective conduction: in the model, this could be observed by changing the dielectric constant or the effective charge of the ions; that is, the difference in the two lines.

The physical meaning of the model result in Fig. 9 is quite significant: it suggests that ion transport in polymer media will be dominated by the coupling to the lattice relaxation, that is, the more rapid the renewal process, the higher the diffusion coefficient. It also suggests, however, that ion correlation can be quite important, and that (in the first instance) these will act to cluster the ions and reduce the overall conduction. On the basis of these models, it is possible to make some important physical suggestions to increase conduction in polymer-based electrolytes. Several of these theoretically based suggestions have been acted on experimentally, and have substantially improved polymer materials. This is a case in which understanding at least a tiny bit of the language of science paid rich dividends in the form of improved materials.

The results of Fig. 9 suggest several important physical themes for increasing electrolyte conductivity and diffusion. First, the dominant effect of renewal suggests that if polymeric electrolytes could be prepared with faster renewal, greater fluidity, lower microviscosity, and therefore higher conduction would result. This can be done by choosing polymers with more flexible backbones and therefore lower glass transition temperatures, plasticized materials, lower salt concentrations (reducing ion induced cross-linking and decreasing glass transition temperatures), and delocalized charge ions. These have all proven to be

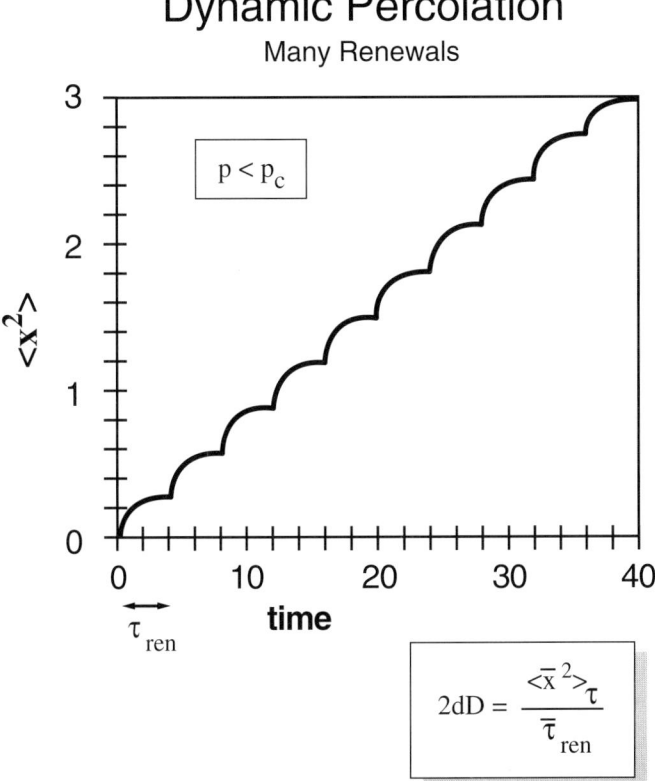

FIG. 7. Result for the mean square displacement as a function of time, for many renewal intervals, for a simple square lattice. On the time scale of the renewal process, the mean square displacement is limited by the size of the connected cluster (the conductivity is below the percolation threshold); for longer times, however, due to the renewal process, the mean square displacement becomes effectively linear in time, thus exhibiting diffusion even below the static threshold. (From ref. [30])

of substantial benefit in improving electrolytes. Indeed, work by Shriver and his colleagues at Northwestern [6] in preparing polyphosphazenes (very low glass transition temperature polymers) provided the polymer electrolytes with highest conductivity at room temperature.

A second implication is simply that reduction of interionic trapping reduces "coulomb drag" on the ions, and increases the conduction. Once again, some straightforward chemistry ideas can be used: large organic ions, in which the charge is delocalized, will ion-pair less strongly than small ones. Molecular en-

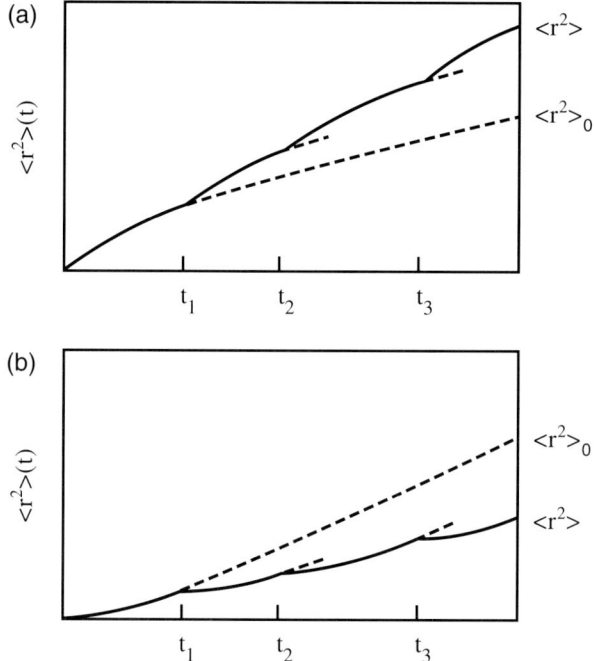

FIG. 8. Mean square displacement as a function of time, for two different physical behaviors. Fig. 8a shows concave downward behavior, corresponding to hopping-like motion (as in polymer electrolytes). Fig. 8b shows a concave up behavior, corresponding to coherent, ballistic transport before the renewal process. For both, motion is diffusive for times substantially exceeding the renewal time. (From Ref. [14])

crypting agents (essential small cyclic ethers, such as shown in Fig. 10, that surround the cation and prevent ion clustering) also increase the conduction, both by minimizing the ion clustering and by decreasing the effective glass transition temperature [11]. Finally, engineering of a polymer whose basic groups (effectively locally negative groups that complex the cation) are of lower basicity yields increased conductivity; this work has produced conductive aluminosilicate-based polymer materials [37].

All of these suggestions are on the chemical end of things, and not necessarily accessible immediately to mathematicians. The important point is that, simply from the modeling results, it is possible to make important statements about synthetic chemistry, statements that have already led to improved materials.

Within the dynamic bond percolation picture, under the assumption of the Nernst–Einstein relationship, a simple reduced dimensionality treatment sug-

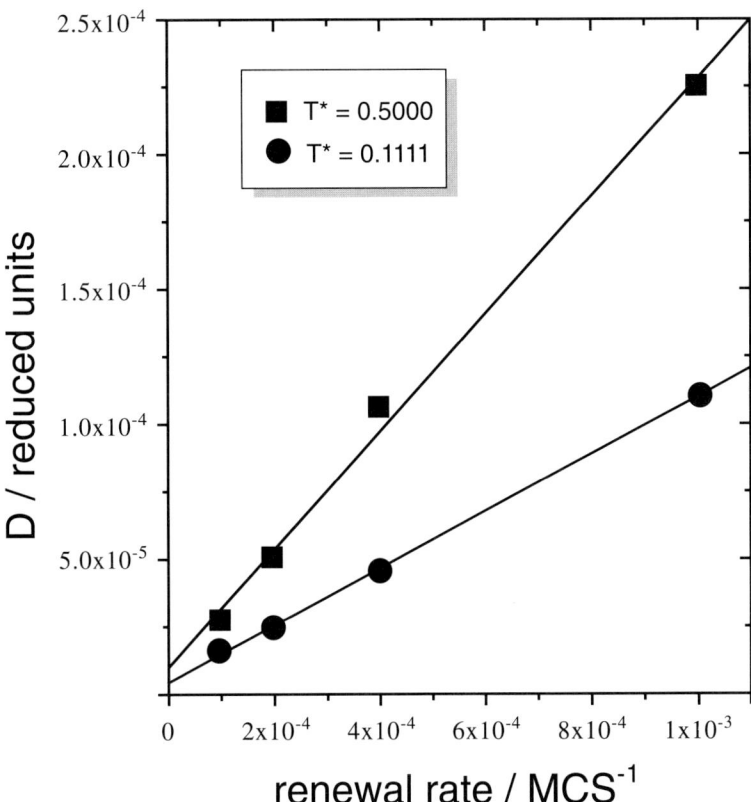

FIG. 9. Diffusion coefficient as a function of renewal frequency, for a three-dimensional dynamically disordered system obeying the rules of dynamic bond percolation theory, but with the charged particles interacting by coulombic forces (with boundary conditions treated appropriately). Abscissa units in μs^{-1}. Physically, the important demonstration here is that diffusion coefficient increases both with increasing renewal frequency and with reduced columbic interaction (stronger screening, higher effective temperature). (From ref. [35])

gests that conductivity is given by [35]:

$$\sigma = Cnq^2 <\overline{r^2}>_0 /(6kT\bar{\tau}\text{ren}). \tag{4.11}$$

Here the symbols σ, C, n, q, and k are respectively the conductivity, a proportionality constant, the number density of carriers, the charge of the carrier, and Boltzmann's constant. By synthesis, one can control the average mean square

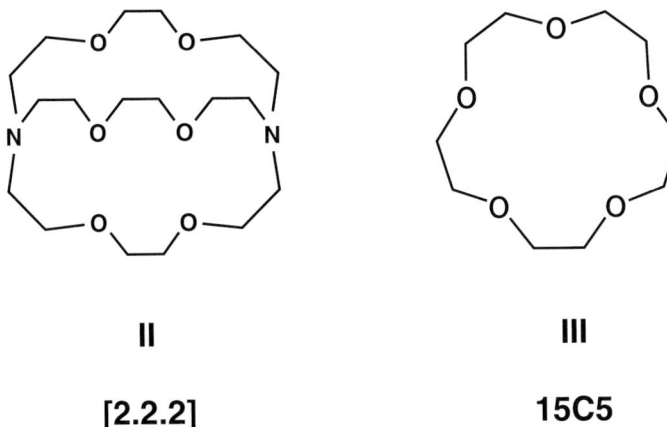

II   **III**

[2.2.2]   **15C5**

FIG. 10. Cyclic ethers, referred to respectively as crown ethers and cryptands. Both of these strongly complex cations such as Na+, reducing ion pairing and increasing conduction in polymer electrolyte materials.

displacement within a given renewal interval (the last term of the numerator of Eq. 4.11), and the renewal time that enters into the denominator. This general equation seems to hold for large classes of polymer electrolyte systems.

The modeling is, of course, neither a substitution for nor a necessary part of chemical synthesis. From the synthetic point of view, however, the results that follow directly from the modeling calculation provide insight into the preparation of new, advanced battery materials.

5 Remarks

This chapter represents more of an impressionistic essay than any hard science; there is some important mathematics involved in the analysis of long-range forces and appropriate boundary conditions, and the topic of renewal theory underlies much of the work on the dynamic percolation picture for polymer electrolytes. The point to be stressed simply is that there are a number of important and vexing problems in modeling and understanding processes in the chemistry of gases, liquids, solids, and disordered systems for which the appropriate mathematical analysis has not been done, and therefore the results are open to suspicion. The case of long-range forces is striking, since various papers on important subjects by well-reputed people are simply incorrect, because the simulations were not done properly. This is an outstanding example of a situation in which the chemists really need help from the mathematicians.

There is one other aspect of the molecular dynamics problem that has exercised physicists, mathematicians, chemists, materials scientists, engineers and other people for a very long time, and which is definitely worthy of new analysis. Physical systems such as liquid water exhibit interesting dynamical behavior on time scales that run, characteristically, over ten to twenty orders of magnitude. For example, ultrafast solvation relaxation in water occurs on roughly a femtosecond time scale, while some characteristic molecular isomerization processes in aqueous media will occur over a time scale of minutes or longer. The standard way that these different time scales are now understood is based on differing physical assumptions and approximations within the different time scale regimes: molecular dynamics for times in the sub-nanosecond range, Langevin or generalized Langevin, or Fokker–Planck dynamics for times in the nanosecond to millisecond regime, and hopping dynamics for times in the microsecond regime and longer. The conditions under which these different approximations become valid are fairly well understood [41], but a more general methodology, that could smoothly bridge this many time scale regime, would be a major contribution in the area.

Many of us remember with dismay the section of our first complex-variable course that dealt with applications in physics: often the instructor was uninterested, the problems were uninteresting, and the wandering from the elegance of the mathematics was noisome. In the research arena, however, physical systems do offer substantial challenge to the mathematician. These challenges are often unaddressed, or addressed in an oblique fashion that is simply not available to the physical scientists involved. Some rather important mathematicians (Newton, Gauss, Laplace, Fourier, Poincaré, Stokes, etc.) have responded to this challenge in the past. Elegant opportunities exist, as demonstrated clearly by some of the very lovely contributions in this book.

Acknowledgements

I am very grateful to Professor Joe Jerome for the opportunity to participate in (and learn from) the conference, and to contribute to this proceedings. I thank Du Shriver, Abe Nitzan, Victor Barcilon, Bob Eisenberg, Zeev Schuss, Steve Druger, John Perram, Simon deLeeuw and Mark Lonergan for substantial help on the problems of ion motion and its simulation. I am grateful to the Materials Research Center of Northwestern University supported by the NSF (Grant #DMR 9120521), to the chemistry division of the NRL, to the ARO (Grant #DAAH 049510304), and to the advanced battery project of Lawrence Berkeley Laboratory for support of this research.

Bibliography

1. Accasina, F. and Fuoss, R. M. (1979). *Electrolytic Conductance*. Wiley.
2. Alder, B. J. and Wainwright, T. E. (1957). *J. Chem. Phys.*, **27**, 1208.
3. Armand, M. B. (1989). *Faraday Disc. Chem. Soc.*, **88**, 65.
4. Bader, J. S., Kuharski, R. A., and Chandler, D. (1990). *J. Chem. Phys.*, **93**, 230.
5. Bader, J. S. and Chandler, D. (1992). *J. Phys. Chem.*, **96**, 6423; Chandler, D. (personal communication).

6. Blonsky, P. M., Shriver, D. F., Austin, P. E., and Allcock, H. R. (1984). *J. Am. Chem. Soc.*, **106**, 6854.
7. Caillol, J. M., Levesque, D., and Weis, J. J. (1989). *J. Chem. Phys.*, **91**, 5544; Caillol, J. M., Levesque, D., Weis, J. J., Patey, G. N. and Kusalik, P. G. (1987). *Mol Phys.*, **62**, 461; Caillol, J. M., Levesque, D. and Weis, J. J. (1986). *J. Chem. Phys.*, **85**, 6645.
8. Caillol, J. M. (1994). *J. Chem. Phys.*, **101**, 6060.
9. Chaikin, P. M. and Lubensky, T. C. (1996). *Principles of Condensed Matter Physics*. Cambridge.
10. Chen, D. P., Eisenberg, R. S., Jerome, J. W., and Shu, C.-W. (1995). *Biophys. J.* **69**, 2304.
11. Chen, K. and Shriver, D. F. (1991). *Chem. Mater.*, **3**, 771.
12. DeLeeuw, S. W., Perram, J. W., and Smith, E. R. (1980). *Proc. Roy. Soc.*, **A373**, 27; DeLeeuw, S. W., Perram, J. W., and Smith, E. R. (1986). *Annu. Rev. Phys. Chem.*, **37**, 245.
13. Druger, S. D., Nitzan, A., and Ratner, M. A. (1983). *J. Chem. Phys.*, **79**, 3133.
14. Druger, S. D. and Ratner, M. A. (1988). *Chem. Phys. Lett.* **151**, 434; Druger, S. D. (1991). *J. Chem. Phys.*, **95**, 2169; Druger, S. D. (1994). *J. Chem. Phys.*, **100**, 3979.
15. Druger, S. D., Ratner, M. A., and Nitzan, A. (1983). *Solid State Ionics*, **9/10**, 1115.
16. Ewald, P. (1921). *Ann. Phys.* (Leipzig), **64**, 253.
17. Gray, F. M. (1991). *Solid Polymer Electrolytes*. VCH, New York.
18. Greenbaum, S. G. (ed.) (1995). *Electrochim. Acta*, **40**, (13/14).
19. Greenbaum, S. G., Wilson, J. J., Wintersgill, M. C., and Fontanella, J. J. (1989). In: *Second International Symposium on Polymer Electrolytes* (ed. B. Scrosati), pp. 35–48. Elsevier.
20. Harrison, A. K. and Zwanzig, R. (1985). *Phys. Rev. A*, **32**, 1072; Hilfer, R. and Orbach, R. (1988). *Chem. Phys.*, **128**, 275; Granek, R. and Nitzan, A. (1992). *J. Chem. Phys.*, **97**, 3823; Loring, R. F. (1991). ibid., **94**, 1505 (A number of related treatments occur in discussion of transport in disordered media; the above are examples.) A recent overview of the DDH model is given by Nitzan, A. and Ratner, M. A. (1994). *J. Phys. Chem.*, **98**, 1765.
21. Hockney, R. W. and Eastwood, J. W. (1981). *Computer Simulation Using Particles*. McGraw-Hill; Allen, M. P. and Tildesley, D. J. (1989). *Computer Simulation of Liquids*. Oxford University Press; Heinzinger, K. (1990). *Computer Modelling of Fluids Polymers and Solids*. Kluwer; Hoover, W. G. (1991). *Numerical Statistical Mechanics*. North Holland; Frenkel, D. and Smit, B. (1996). *Understanding Molecular Simulations*. Academic Press.
22. Jayasinghe, G. D. L. K., Dissanayake, M. A. K. L., Careem, M. A., and

Souquet, J. L. (1997). *Sol. St. Ionics*, **93**, 291; Cheradame, H. (1982). In: *IUPAC Macro-Molecules* (eds. H. Benoit and P. Rempp), p. 351. Pergamon.
23. Jerome, J. and Shu, C.-W. (1995). *IEEE Trans. Comp. Aided Design*, **14**, 917.
24. Jorgensen, W. L., Buckner, J. K., Huston, S. E., and Rossky, P. J. (1989). *J. Am. Chem. Soc.*, **109**, 1891 (cf. also Ref. [32]).
25. Koksbang, R. *et al.* (1994). *Sol. St. Ionics*, **69**, 320.
26. MacCallum, J. R. and Vincent, C. A. (1987). *Polymer Electrolyte Reviews 1*. Elsevier; MacCallum, J. R. and Vincent, C. A. (1989). *Polymer Electrolyte Reviews 2*. Elsevier.
27. Montroll, E. W. and Weiss, G. H. (1965). *J. Math. Phys.*, **6**, 167; Montroll, E. W. (1969). *J. Math. Phys.*, **18**, D53; Feller, R. W. (1971). *An Introduction to Probability Theory*, Vol. 2. Wiley.
28. Müller-Plathe, F. (1994). *Acta Polymer*, **45**, 259; Neyertz, S., Brown, D. and Thomas, J. O. (1994). *J. Chem. Phys.*, **101**, 10064; Negerts, S. and Brown, D. (1996). *J. Chem. Phys.*, **104**, 3797; *Comput. Polymer Sci.* (in press); Catlow, C. R. A. and Mills, G. E. (1994). *J. Chem. Soc. Chem. Comm.*, **18**, 2037.
29. National Academy of Sciences (1995). *Mathematical Challenges from Computational Chemistry*. National Academy Press, Washington.
30. Neyertz, S., Brown, D. and Thomas, J. O. (1995). *Electrochim. Acta*, **40**, 2063.
31. Onsager, Lars (1996). *[Collected Works]* (ed. P. C. Hemmer *et al.*). World Scientific.
32. Payne, V.A., Xu, J.-H., Forsyth, M., Ratner, R.A., Shriver, D.F., and deLeeuw, S.W. (1995). *J. Chem. Phys.*, **103**, 8734; Payne, V. A., Forsyth, M., Ratner, M. A., Shriver, D. F., and deLeeuw, S. W., (1994). *J. Chem. Phys.*, **100**, 5201; Forsyth, M., Payne, V. A., Ratner, M. A., deLeeuw, S. W., and Shriver, D.F. (1992). *Solid State Ionics*, **53/56**, 1011; Payne, V. A., Xu, J.-H., Forsyth, M., Ratner, M. A., and Shriver, D. F. (1995). *J. Chem. Phys.*, **103**, 8746; Payne, V. A., Lonergan, M. C., Forsyth, M., Ratner, M. A., Shriver, D. F., deLeeuw, S. W., and Perram, J. W. (1995). *Solid State Ionics*, **81**, 171; Payne, V. A., Xu, J.-H., Forsyth, M., Ratner, R. A., Shriver, D. F., and deLeeuw, S. W. (1995). *Electrochim. Acta.*, **40**, 2087.
33. Perera, L., Essmann, U., and Berkowitz, M. (1995). *J. Chem. Phys.*, **102**, 450.
34. Rahman, A. (1964). *Phys. Rev.*, **136A**, 405.
35. Ratner, M. A. and Nitzan, A. (1989). *Discuss. Faraday Soc.*, **88**, 19; Lonergan, M. C., Nitzan, A., Ratner, M. A., and Shriver, D. F. (1995). *J. Chem. Phys.*, **103**, 3253; Lonergan, M. C., Perram, J. W., Ratner, M. A., and Shriver, D. F. (1993). *J. Chem. Phys.*, **98**, 4937; Lonergan, M. C. (1994).

Ph.D. thesis, Northwestern University, Evanston.
36. Ratner, M. A. and Shriver, D. F. (1988). *Chem. Rev.*, **88**, 109.
37. Rawsky, G. C., Henrietta, K. J., Lowrey, R., Shriver, D. F., and Vayman, S. (1995). *Mat. Res. Soc. Symp. Proc.*, **393**, 189.
38. Roberts, J. E. and Schnitker, J. (1995). *J. Phys. Chem.*, **99**, 1322.
39. Saito, M. (1994). *J. Chem. Phys.*, **101**, 4055; Huston, S. E. and Rossky, P. J. (1989). *J. Phys. Chem.*, **73**, 7888.
40. Schatz, G. C. and Ratner, M. A. (1994). *Quantum Mechanics in Chemistry.* Prentice-Hall.
41. Schuss, Z. (1980). *Theory and Applications of Stochastic Differential Equations.* Wiley; Gardiner, C. W. (1983). *Handbook of Stochastic Methods.* Springer; Wax, N. (ed.) (1954). *Selected Papers on Noise and Stochastic Processes.* Dover.
42. Smedley, S. I. (1980). *The Interpretation of Ionic Conductivity in Liquids.* Plenum Press.
43. Smith, P. E. and Pettitt, B. M. (1991). *J. Chem. Phys.*, **95**, 8430.
44. Stauffer, D. and Aharony, A. (1992). *Introduction to Percolation Theory.* Taylor and Francis, London.
45. Thomas, J. O. (ed.) (1997). *Proc. Ninth Symp. Polymer Electrolytes.* Elsevier.
46. Tonge, J. S. and Shriver, D. F. (1989), In: *Polymers for Electronic Applications* (ed. J. H. Lai), vol. 157. CRC, Boca Raton.
47. Tunaley, J. K. E. (1974). *Phys. Rev. Lett.*, **33**, 1037; Haus, J. W., Kehr, K. W., and Lyklema, J. W. (1982). *Phys. Rev. B*, **25**, 2905; Scher, H. and Lax, M. (1973). *Phys. Rev. B*, **7**, 4502.
48. Vineyard, G.H. (1957). *J. Phys. Chem. Solids*, **3**, 121.
49. Wintersgill, M. C., Fontanella, J. J., Calame, J. P., Figueroa, D. R., and Andeen, C. G. (1983). *Solid State Ionics*, **11**, 151; Wintersgill, M. C., Fontanella, J. J., Greenbaum, S. G., and Adamic, K. J. (1988). *Brit. Polym. J.*, **20**, 195.

2
Transport of Multispecies Contaminants with Biological and Chemical Kinetics in Porous Media

Mary F. Wheeler, Clint Dawson, and Joe Eaton
The University of Texas at Austin

1 Introduction

The contamination of groundwater is a very serious environmental problem. Over half of the U.S. population depends on groundwater for its water supply. Available sources of groundwater are a fundamental constraint on the development and economic activity in many regions.

The characterization and remediation of contaminated sites is difficult and expensive and only now is technology emerging to cope with this severe and widespread problem. Modeling of multiphase flow in permeable media plays a central role in these technologies and is essential for risk assessment, cost reduction, and the rational and effective use of resources. New machine and algorithmic developments can benefit the understanding, design, and testing of economically feasible decontamination strategies. In particular, parallel simulators have the potential to solve larger, more realistic problems then previously possible. In today's information technology explosion there is much interest in incorporating advanced, scalable distributive memory (parallel) algorithms and multiscale nonlinear and stochastic science into a problem-solving environment in order to resolve the physical complexities of the governing nonlinear equations.

In this chapter we will briefly describe some of the results obtained by the Center for Subsurface Modeling (CSM) at The University of Texas at Austin. This group comprises a close-knit team of faculty and research scientists with expertise in applied mathematics, engineering, physical, chemical, and geological sciences, and computer science and uses a multidisciplinary approach to the investigation of high-performance parallel processing as a tool to model the behavior of fluids in permeable geological formations.

The major thrust of this chapter is to describe the general structure of the mathematical and physical models required for the simulation of subsurface flow and transport in heterogeneous porous media and the scalable numerical algorithms and computer science infrastructure necessary for accurate and efficient simulation. The outline of the chapter is as follows. In Section 2, we present a model formulation for multicomponent, multiphase flow that includes mass transfer, geochemical, and biochemical effects. To illustrate the coupling of fluid phase behavior with flow and transport in a simple fashion, a section on the black-oil model is given. In Section 3, we discuss some of the mathematical and algorithmic issues arising in the numerical solution of subsurface flow equations. In Section 4, we discuss the outline of a problem-solving environment for modeling flow in permeable media. In the last section we present results on a problem involving flow and transport biochemical kinetics in a heterogeneous porous medium.

2 Model Formulation

We shall consider the following multicomponent, multispecies, multiphase flow model which includes equilibrium and nonequilibrium mass transfer between phases, geochemistry, biochemistry, and radionuclide decay. Here we use the terms *components* and *species*, where a component refers to a basic chemical entity such that every species can be uniquely represented as a combination of components. A species is a product of a chemical reaction involving components as reactants.

Let n_c denote the number of components in the system. Let n_p denote the total number of flowing and stationary phases in the system, and n_s the maximum number of species in the system.

The flow of species k within phase j can be written as [11, 10]

$$\frac{\partial N_{kj}}{\partial t} + \nabla \cdot \mathbf{F}_{kj} = R_{kj}, \tag{2.1}$$

where N_{kj} is the molar density (moles/unit volume of phase j), $\mathbf{F}_{kj}$ is the species flux, and R_{kj} describes kinetic source and sink terms.

The species flux can be written as

$$\mathbf{F}_{kj} = \xi_j x_{kj} \mathbf{u}_j - \phi \xi_j S_j \mathbf{K}_{kj} \nabla x_{kj}, \tag{2.2}$$

where ξ_j is the molar density of phase j, x_{kj} is the mole fraction of species k in phase j, $\mathbf{u}_j$ is the Darcy velocity of phase j, ϕ is the porosity of the medium, S_j is the saturation of phase j, and $\mathbf{K}_{kj}$ is the diffusion/dispersion tensor for the species. We note that $\xi_j x_{kj} = N_{kj}$, $\sum_{k=1}^{n_s} x_{kj} = 1$, and $\sum_j S_j = 1$. We also note that (2.1) does not preclude changes in the porosity of the media due to, for example, biomass growth.

The flowing phase velocity is given by Darcy's Law:

$$\mathbf{u}_j = -\frac{K k_{rj}}{\mu_j} \left(\nabla P_j - \rho_j g \nabla D \right), \tag{2.3}$$

where k_{rj} is the relative permeability, μ_j is the viscosity, P_j is the pressure and ρ_j is the mass density for phase j, and $g \nabla D$ is a gravitational force vector. Here

$$\frac{K k_{rj}}{\mu_j} \equiv \lambda_j = \lambda_j(S, x, \mathbf{x}, T), \tag{2.4}$$

where S is a vector of phase saturations, x is a vector of species mole fractions, $\mathbf{x}$ is spatial location, and T is temperature. The capillary pressure relation gives

$$P_j - P_n = P_{cjn}(S, x, \mathbf{x}, T), \quad j \neq n. \tag{2.5}$$

This relation may also be hysteretic. Molar and mass densities, viscosity, and porosity are assumed to be functions of composition, pressure, and temperature, with porosity also varying spatially:

$$\xi_j = \xi_j(x, P_j, T), \quad \rho_j = \rho_j(x, P_j, T), \tag{2.6}$$
$$\mu_j = \mu_j(x, P_j, T), \quad \phi = \phi(\mathbf{x}, x, P_j, T). \tag{2.7}$$

For nonmobile species (such as microorganisms attached to the pore surface), the flux term $\mathbf{F}_{kj} = 0$.

The reaction term R_{kj} can be quite complicated, involving numerous types of reactions. In particular,

$$\begin{aligned} R_{kj} = {} & \text{mass transfer of species } k \text{ into} \\ & \text{and out of phase } j \\ & + \text{ geochemical reactions} \\ & + \text{ biochemical reactions} \\ & + \text{ radionuclide decay} \\ & + \text{ external sources and sinks.} \end{aligned} \tag{2.8}$$

The geochemical reactions include reactions within the phase (complexation and redox), sorption (adsorption and ion exchange) and precipitation/dissolution.

In many cases of interest, the reactions occur at such fast rates relative to transport that they are assumed to be in local equilibrium. In the case of equilibrium mass transfer, the species balance equations are generally summed over phases (noting that the mass transfer terms sum to zero), giving an equation for the total species molar density:

$$\frac{\partial N_k}{\partial t} + \sum_{j=1}^{n_p} \nabla \cdot F_{kj} = \sum_{j=1}^{n_p} R_{kj}. \tag{2.9}$$

In practice, one solves for the total species by some means and performs a "flash calculation" to determine the mole fraction of the species within each phase.

When modeling geochemistry, similar manipulations can be performed to form equations for total component densities. Multiplying (2.1) by stoichiometric coefficients a_{ik}, determined by

$$\sum_{k=1}^{n_s} a_{ik} N_{kj} = \tilde{N}_{ij}, \tag{2.10}$$

where $\tilde{N}_{ij}$ is the molar density of component i in phase j, summing on species and phases, we obtain an equation for the total molar density of the component:

$$\frac{\partial \tilde{N}_i}{\partial t} + \sum_{k=1}^{n_s} a_{ik} \sum_{j=1}^{n_p} \nabla \cdot F_{kj} = \sum_{k=1}^{n_s} a_{ik} \sum_{j=1}^{n_p} R_{kj}. \tag{2.11}$$

Since the total reaction rate must sum to zero, the right hand side in (2.11) reduces to a sum of external sources and sinks and radionuclide decay terms only.

For both equilibrium mass transfer and geochemistry, determining mole fractions of each species in each phase involves minimizing the Gibbs free energy, subject to constraints. Equivalent methods are available for specialized forms of phase equilibrium such as needed for microemulsions [6]. For mass transfer, simplified methods such as Henry's Law are often used for dilute solutions or trace components

Nonequilibrium reactions require returning to (2.1) to determine the mole fraction of a species within a phase. A functional form for the reaction term must be specified. For nonequilibrium mass transfer, for example, the reaction term which would appear on the right side of (2.1) is generally of the form

$$\kappa_k \left(x_{kj}^{eq} - x_{kj} \right), \tag{2.12}$$

where κ_k is an effective mass transfer coefficient and x_{kj}^{eq} is obtained through an equilibrium calculation [17, 12, 16]. In chemical reactions, a similar though more complicated expression is commonly used. Thus nonequilibrium reactions may also require solving a constrained minimization problem.

In order to complete the model, we need an energy balance. The molar energy balance for the control volume using internal energy rather than temperature as a primary variable, assuming local equilibrium heat transfer between phases, can be expressed as [4]

$$\frac{\partial u}{\partial t} + \nabla \cdot \sum_{j=1}^{n_p} \xi_j h_j \mathbf{u}_j - \nabla \cdot (\lambda_T \nabla T) - q_H + q_L = 0. \tag{2.13}$$

Here $u = u_r + u_f$ is the sum of the internal energies of the minerals (rock or soil) and the fluids per unit bulk volume, h_j denotes the molar enthalpy of phase j, λ_T is the effective thermal conductivity, q_H is the enthalpy injection rate per unit bulk volume, and q_L is heat loss to the over-and underburdens per unit bulk volume.

In summary, equations for total species molar density or total component molar density can be formed using (2.9) or (2.11). Within these equations are mole fractions of species within a phase. These mole fractions are determined by an equilibrium relationship or the nonequilibrium equation (2.1). These equations, together with Darcy's Law, the constitutive relationships (2.4)–(2.7), volume balance, energy balance, and appropriate boundary and initial conditions, comprise the mathematical model.

2.1 Black-Oil Model

As a special case of the general transport model described above, we present here in some detail the basic equations corresponding to the black-oil model, which has been widely used in the petroleum industry.

In the black-oil model, it is assumed that there are only three components and three phases, both denoted by oil, gas, and water. All fluids are at constant temperature and in thermodynamic equilibrium throughout the reservoir. Furthermore, mass transfer between phases is allowed in the following way: the gas component can exist in all three phases, the oil component can be present in the oil and gas phases, and the water component only exists in the water phase.

For each component, the mass conservation equation can be written as

$$\frac{\partial}{\partial t}[\phi N_I] + \nabla \cdot \left[\sum_{J=1}^{N_P} C_{IJ}\, \rho_J\, \mathbf{u}_J\right] + q_I = 0, \tag{2.14}$$

where ϕ is porosity, N_I is the total mass density of component I, with $I = O, G, W$ for oil, gas, and water, respectively. The mass fraction coefficients C_{IJ} represent the amount of component I in phase J, and will have to be determined by phase equilibrium as described below. They are related to the component densities by

$$N_I = \sum_{J=1}^{N_P} C_{IJ}\, S_J\, \rho_J$$

where S_J are phase saturations and ρ_J phase densities. Finally, q_I represents a source term for component I.

In addition to the three mass conservation equations for each component, we have a volume constraint so that the fluids fill the pore volume:

$$S_o + S_w + S_g = 1. \tag{2.15}$$

(Following standard notation, we use capital letter subscripts for component variables ($I = O, G, W$) and lower-case letter subscripts for phase variables ($J = o, g, w$).) Moreover, capillary pressures are assumed in the form

$$p_{Cow} = p_o - p_w = f(S_o, S_g, S_w), \tag{2.16}$$

$$p_{Cog} = p_g - p_o = f(S_o, S_g, S_w). \tag{2.17}$$

Additional relations for relative permeabilities $k_{rJ} = k_{rJ}(S_o, S_g, S_w)$ and phase viscosities $\mu_J = \mu_J(p_J, C_{IJ})$ complete the system of equations.

Thus, the flow and transport problem consists of six equations (three equations as given by (2.1), plus (2.2)–(2.4)) with six unknowns: three mass components N_O, N_G, N_W and three phase pressures p_o, p_g, p_w. By use of (2.3)–(2.4), the system can be reduced to 4×4 with primary variables given by, for instance, N_O, N_G, N_W and p_w.

Coupled with the flow description, there is the phase equilibrium problem that governs the thermodynamic conditions under which the components combine to form phases. By introducing certain functions of pressure and temperature obtained through experimental measurements, it is possible to obtain mass fractions C_{IJ}, phase densities ρ_J, and saturations S_J given a state with known primary variables N_O, N_G, N_W, and p_w. For example, when all three phases are present, the component densities and saturations are related through the following expression [15]:

$$\begin{bmatrix} N_O/N_{OS} \\ N_G/N_{GS} \\ N_W/N_{WS} \end{bmatrix} = \begin{bmatrix} B_o^{-1} & R_v B_g^{-1} & 0 \\ R_{so} B_o^{-1} & B_g^{-1} & R_{sw} B_w^{-1} \\ 0 & 0 & B_w^{-1} \end{bmatrix} \begin{bmatrix} S_o \\ S_g \\ S_w \end{bmatrix},$$

where the subscript S in the component densities denotes standard conditions. The functions B_o, B_g and B_w are the formation volume factors for each phase. R_{so} defines the solubility of gas in oil, R_{sw} is the solubility of gas in water, and R_v gives the volatility of oil in gas. All these quantities are known functions of pressure and temperature. Similar equations can be derived for the mass fractions C_{IJ} and phase densities ρ_J.

It should be noted that a thermodynamic test is required to define the number of phases present at a given state. In correspondence with our mass transfer assumptions, it is possible to have either the gas phase missing or the oil phase missing. In the first case, all gas is dissolved in oil and/or water (undersaturated oil), while in the second case, all oil is dissolved into gas (undersaturated gas). For these cases, slightly different equations have to be employed in the phase equilibrium problem [20]. Notice, however, that the flow equations (2.14) are still valid when one phase is missing. This is why we use the component densities N_I as primary variables instead of saturations S_J, which would appear to be a better choice in view of the volume constraint (2.15), but would require to modify the flow equations as one phase disappears.

3 Mathematical and Algorithmic Issues

We remark on the scaling in porous media problems. The process scale varies from 10^{-6} to 10^{-3} m^3 and the field scale is 10^7 to 10^{12} m^3. This implies that 10^{10} to 10^{12} elements are needed for a physically representative solution. Presently, 10^6 elements are considered cutting-edge. In addition, because of the uncertainties in the coefficients, hundreds of simulations are needed for accurately quantifying prediction.

From this overview, it is clear that the simulation of complex subsurface phenomena on multiple realizations requires a flexible yet extremely powerful computational approach. To cope with this challenge, over the last years two main modeling approaches have emerged, which in the near future will be tied into a single problem-solving environment.

The first modeling approach emphasizes computational speed at the expense of model complexity. This set of simulators is based on a sequential decoupling of the flow and reactive transport processes. Problems involving two and three flowing phases with weakly compositional phase behavior are solved within this framework (i.e., dilute mixtures and low-pressure applications).

The second modeling approach is appropriate for problems with a tight coupling between flow and reactive transport. It requires a fully compositional formulation for phase behavior, with great demands in computational resources. Problems with complex phase behavior and high pressure fields fall into this group, such as petroleum and groundwater spill and non-aqueous phase liquid (NAPL) applications.

Researchers at the Center for Subsurface Modeling in collaboration with researchers at the Center for Petroleum and Geosystems at The University of Texas at Austin have been working on both approaches as well as in the implementation of a unified problem-solving environment which will be discussed in a later section. Particular emphasis has been placed on the following areas:
- numerical algorithms, in particular for tightly coupled multicomponent, multiphase flow models;
- parallel linear and nonlinear solvers, which are crucial for the overall efficiency and robustness of the simulators;
- efficient techniques for solving the phase behavior problem, which in its most general form involves the minimization of Gibbs free energy;
- geostatistical modeling and scale-up techniques for uncertainty estimation and risk assessment;
- incorporation of equilibrium and nonequilibrium geochemical reactions and biodegradation modules into the flow model.

In the next subsection we briefly discuss some of the algorithmic approaches we have found promising in addressing the first three areas.

3.1 Algorithms

In modeling flow in porous media, it is important to employ a discretization method that is accurate, conserves mass locally, and preserves continuity of the phase fluxes. In addition, the algorithm needs to be able to handle tensor permeabilities and irregularly shaped domains. We have chosen to use mixed finite element methods which satisfy all of the above properties. In particular they conserve mass locally. For transport, we have investigated both characteristic and higher-order Godunov methods for modeling advection, combined with the mixed finite element method for modeling diffusion. These methods are also conservative.

For mixed finite element approximations to elliptic partial differential equations, Arbogast, Cowsar, Wheeler, and Yotov [1, 3] recently introduced mortar spaces on interfaces for multiblock domains. In the case of the lowest-order Raviart Thomas spaces on logically rectangular grids for each subdomain, superconvergence was established. Moreover, efficiency was not sacrificed by adding the mortar unknowns. The computational complexity was shown to be comparable to the one on matching grids. Numerical results have verified the theory.

This work can be generalized as was done by Yotov [22] to two-phase problems. Other extensions include application of non-matching grids with different approximations and different physics in different domains.

To address the solution of the nonlinear equations arising in a fully implicit formulation, the hybrid Krylov-secant methods were formulated [8, 9, 7]. Here a novel way of reusing the Krylov information generated by GMRES within an inexact Newton method has been introduced. The approach is motivated by theory recently developed on secant preconditioners and combines Broyden updates of the Hessenberg matrix generated by the Arnoldi process in GMRES, the Richardson iteration, and limited memory quasi-Newton compact representations to generate descent directions for each Newton step. This method incorporates secant updates of the Jacobian and its preconditioner and uses spectral information stemming from the Arnoldi factorization to optimize the efficiency of a cheaper nonstationary Richardson iteration. Hence the resulting method turns out to be computationally more economical than traditional inexact Newton implementations that rely upon successive iterative solutions from scratch of the Jacobian linear system. This procedure has been successfully applied to two-phase problems with highly heterogeneous permeability fields and multiple wells.

The computation of thermodynamic equilibrium is a key subproblem in the simulation of environmental problems involving reactive transport. The formulation of the equilibrium problem as a constrained minimization of the Gibbs free energy offers certain advantages over a mass-action formulation, especially for applications in which phases may appear and disappear. In [19, 18], the interior point method was introduced for computing general chemical equilibria. This approach is attractive in that it does not require special treatment of any phase. Furthermore, it conveniently incorporates inequality constraints, which necessarily arise on physical grounds in these problems.

For modeling advection-dominated transport, we have considered two methods, one based on characteristic methods and one based on higher-order Godunov methods. The characteristic-mixed method was developed by Arbogast and Wheeler [2] and employs characteristic traceback of volumes to determine the advective transport. The Godunov-mixed method was developed by Dawson [5]; here, advective fluxes through each grid block edge are approximated by characteristic traceback. In both methods, discontinuous, piecewise linear approximations are used and diffusion/dispersion is incorporated by the mixed finite element method. Both methods exhibit second-order convergence in space and first-order convergence in time. With minor modifications, the Godunov-mixed method can be made second order in time.

4 A Problem-Solving Environment for Flow in Permeable Media

Key requirements in the development of a problem-solving environment (PSE) for porous media applications include the support of high-resolution studies with millions of grid elements, ability to handle multiple physical models, multiple fault blocks, well management, dynamic locally adaptive mesh refinements, and interactive visualization and computational steering.

The main objective of the PSE is to reduce the complexity of building flexible and efficient parallel porous media simulators through the use of a high-level programming interface for problem specification and model composition, object-oriented programming abstractions that implement application objects, and distributed dynamic data management that efficiently supports adaptation and parallelism. Secondary objectives include developing a general framework for integrating input/output, visualization, and interactive experimentations with applications computation, and to use the most appropriate technology for individual tasks in the PSE. For example, to implement low-level data management in C, to build programming abstractions in object-oriented language like C++, and to develop computational kernels in FORTRAN.

At CSM, the PSE design being implemented is based on a clear separation of concerns across its functionality and the definition of hierarchical levels of abstraction based on this separation. Its compact and efficient implementation draws upon the capabilities provided in object-oriented programming systems. Each layer is implemented as a set of C++ classes which inherit from the classes defined in the lower layers or use composition/template mechanisms to create container classes. The lowest layer of the PSE is a Hierarchical Distributed Dynamic Array (HDDA). HDDA provides pure array semantics to hierarchical, dynamic and physically distributed data, and has been successfully used as a data-management infrastructure for a variety of parallel adaptive algorithms [13]. The next layer adds application semantics to HDDA objects and implements application objects such as grids, meshes, and trees. This layer provides object-oriented programming abstractions that can be used to directly implement parallel adaptive algorithms. The upper layers of the PSE implement application specific methods and components. The topmost layer is a high-level application programming interface customized to parallel porous media simulation. It provides keyword input support for problem and model specification, generalized units conversion, and incorporates a FORTRAN interpreter to allow code fragments as part of the user input specification. More detailed discussion of this PSE can be found in [21, 14].

5 Numerical Results

Below we present numerical results for a biogeochemical remediation study. In this study we assume electron donors (contaminants) and microorganisms are in place. Two cases are considered, based on activating a microbial pollution breakdown via oxygen injection upstream. In the first case we assume a passive porous medium; i.e., the transporting species do not interact with the porous medium. In the second case, we incorporate reducing mineral components in the medium which retard the transport of oxygen.

In these simulations, the polluted region is assumed to be a cube. A $32 \times 32 \times 32$ grid for pressures and concentrations was employed. Porosity was assumed to be 0.30. A hydraulic conductivity field was supplied from real field data.

The CSM Parssim1 simulator was used to conduct these experiments. Here, the flow was approximated by a mixed finite element scheme. Transport was handled by by the higher-order Godunov scheme described above. General chemistry, in which equilibrium and rate-limited reactions were simultaneously modeled, was performed using an interior point Newton method for minimizing Gibbs free energy. The chemical reactions were modeled in a split step via operator splitting.

The passive case uses second components to model rate-limited nonaqueous phase liquid (NAPL) dissolution in a saturated media coupled with aerobe biodegradation of the dissolved NAPL over a time span of five years. The aerobes are activated by injecting oxygenated fluid upstream of the contaminated site. When both oxygen and dissolved NAPL are present, the aerobe population increases, accelerating the rate of degradation into harmless by-products. Control of downstream contamination is achieved in 1.5 model years in this case. A plot of the contaminant concentration at 1 year of the study is given in Fig. 1. The white regions in the figure illuminate the areas with highest concentration.

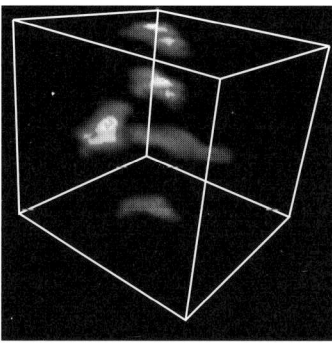

FIG. 1. Contaminant concentration at one year, passive media

When a reducing agent is added to the media, the model requires 9 components. The arrival time of the oxygenated fluid is delayed by a significant amount. Oxygen is consumed in the reduction reaction before it can reach the aerobes and activate the degradation reactions. Downstream pollution is thus much greater in this model, and control is not attained until 2.5 years after injection begins. In Fig. 2, we again show the contaminant concentration at 1 year.

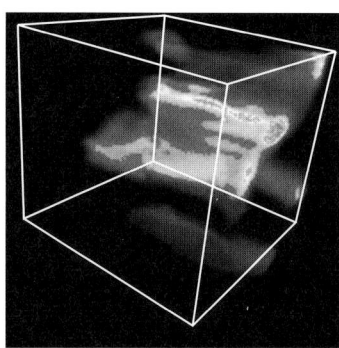

FIG. 2. Contaminant concentration at one year, reacting media

These cases demonstrate that neglecting the geochemistry when modeling soil contamination can lead to inaccurate and misleading results and possibly encourage bad decision-making regarding treatment of contamination.

Acknowledgements

The development of a problem-solving environment project at CSM project involves an interdisciplinary group of researchers with considerable experience in parallel computation. The applied mathematics and engineering component, Professors Mary Wheeler, Todd Arbogast, Clint Dawson, and Drs. Steve Bryant and Ivan Yotov have developed massively parallel numerical algorithms for the modeling of multiphase subsurface flow, transport, and geo- and biochemical reactions. The computer science component, Professor J.C. Browne and Dr. Manish Parashar, have worked in formulating a general problem-solving framework for dynamic mesh refinements and have developed the software package Dynamic Adaptive Grid Hierarchy (DAGH).

We also acknowledge BP Exploration for providing the permeability field used in the numerical simulations above.

Bibliography

1. Arbogast, T., Cowsar, L., Wheeler, M.F. and Yotov, I. (1996). Mixed finite element methods on non-matching grids (submitted for publication).
2. Arbogast, T. and Wheeler, M.F. (1995). A characteristic-mixed finite element method for advection-dominated transport problems. *SIAM J. Numer. Anal.*, **32**, 404–424.

3. Arbogast, T., Wheeler, M.F. and Yotov, I. (1996). Logically rectangular mixed methods for flow in irregular, heterogeneous domains. In: *Computational Methods in Subsurface Flow and Transport Problems* (eds. A. A. Aldama, J. Aparicio, C. A. Brebbia, W. G. Gray, I. Herrera, and G. F. Pinder), vol. I, pp. 621–629. Computational Mechanics Publications, Southampton, UK.
4. Brantferger, K., Pope, G. A., and Sepehrnoori, K. (1991). Development of a thermodynamically consistent, fully implicit, compositional, equation-of-state steamflood simulator, In: paper SPE 21253, presented at the 11th *SPE Symposium on Reservoir Simulation*, Anaheim, CA.
5. Dawson, C. N. (1993). Godunov-mixed methods for advection-diffusion equations in multidimensions. *SIAM J. Numer. Anal.*, **30**, 1315–1332.
6. Delshad, M., Pope, G. A., and Sepehrnoori, K. (1997). A compositional simulator for modeling surfactant enhanced aquifer remediation 1: Formulation. *J. Contaminant Hydrology*, to appear.
7. Klíe, H. (1996). Krylov-secant methods for solving large scale systems of coupled nonlinear equations. Ph.D. thesis, Rice University.
8. Klíe, H., Ramé, M., and Wheeler, M. (1995). Krylov–secant methods for solving systems of nonlinear equations. *Tech. Rep. TR95-27*, Dept. of Computational and Applied Mathematics, Rice University, 1995.
9. Klíe, H., Ramé, M., and Wheeler, M. (1995). Hybrid Krylov-secant methods for nonlinear equations arising in a porous media application. In: *Computational Methods in Subsurface Flow and Transport Problems* (eds. A. A. Aldama, J. Aparicio, C. A. Brebbia, W. G. Gray, I. Herrera, and G. F. Pinder), vol. I, pp. 467–481. Computational Mechanics Publications, Southampton, UK.
10. Lake, L. W. (1989). *Enhanced Oil Recovery*. Prentice Hall.
11. Lake, L. W., Pope, G. A., Carey, G. F., and Sepehrnoori, K. (1984). Isothermal, multiphase, multicomponent fluid flow in permeable media, *In Situ*, **8**, 1–40.
12. Miller, C. T., Poirier-McNeill, M. M., and Mayer, A. S. (1990). Dissolution of trapped nonaqueous phase liquids: Mass transfer characteristics. *Water Resources Research*, **26**, 2783–2796.
13. Parashar, M. and Browne, J. C. (1996). Object oriented programming abstractions for parallel adaptive mesh refinement. In: *Parallel Object-Oriented Methods and Applications Workshop*, Santa Fe, NM.
14. Parashar, M., Wheeler, J. A., Pope, G. A., Wang, K., and Wang, P. (1997). A new generation eos compositional reservoir simulator: Part ii–framework and multiprocessing. In: paper SPE 37977, presented at the *SPE Symposium on Reservoir Simulation*, Dallas, TX.
15. Peaceman, D. W.(1977). *Fundamentals of Numerical Reservoir Simulation*. Elsevier.

16. Powers, S. E., Abriola, L. M., and Weber, J. W. J.(1994). An experimental investigation of nonaqueous phase liquid dissolution in saturated subsurface systems: Transient mass transfer rates. *Water Resources Research*, **30**, 321–332.
17. Powers, S. E., Louriero, C. O., Abriola, L. M. and Weber, J. W. J.(1991). Theoretical study of the significance of nonequilibrium dissolution of nonaqueous phase liquids in subsurface systems. *Water Resources Research*, **27**, 463–477.
18. Saaf, F. (1996). A study of reactive transport phenomena in porous media. Ph.D. thesis, Rice University.
19. Saaf, F., Tapia, R., Bryant, S., and Wheeler, M. F.(1996). Computing general chemical equilibria with an interior point method. In: *Computational Methods in Subsurface Flow and Transport Problems* (eds. A. A. Aldama, J. Aparicio, C. A. Brebbia, W. G. Gray, I. Herrera, and G. F. Pinder), vol. I, pp. 201–209. Computational Mechanics Publications, Southampton, UK.
20. Trangestein, J. A. and Bell, J. B.(1989). Mathematical structure of the black-oil model for petroleum reservoir simulation. *SIAM J. Appl. Math.*, **49**, 749–783.
21. Wang, P., Yotov, I., Wheeler, M. F., Arbogast, T., Dawson, C., Parashar, M., and Sepehrnoori, K. (1997). A new generation eos compositional reservoir simulator: Part i–formulation and discretization, In: paper SPE 37977, presented at the *SPE Symposium on Reservoir Simulation*, Dallas, TX.
22. Yotov, I. (1996). Mixed finite element methods for flow in porous media. Ph.D. thesis, Rice University.

3

Equidistribution and Extremal Energy of N Points on the Sphere

Y. M. Zhou
Northwestern University

Abstract

In this chapter, we will investigate the problem of arrangement of N points on the surface of the unit sphere $\mathcal{S}^2$ in $\mathbb{R}^3$, so that the associated potential energy is extremal. We will also discuss the relationship exhibited between the Dirichlet cell structure of the numerical solution of the extremal energy problem and the structure of carbon fullerenes and the possible application directed toward new carbon fullerene structures. An explicit formula that generates any number of points, called Generalized Spiral Points, evenly spaced on the sphere, is presented. We address the problem of mild concentration of the spiral points near the poles. We also introduce a variant procedure, and present a c-code segment for obtaining N points evenly spaced on the sphere.

1 Introduction

Circles and spheres have intrinsic properties that continue to fascinate mathematicians and scientists. Problems of arranging points on the sphere arise naturally in biology, chemistry, and physics. Ever since J. J. Thomson's plum pudding model of the atom, there has been much interest in the equilibrium configurations of electrons confined to spheres and disks. Although the original motivation was quickly made obsolete by the advent of quantum mechanics, interest in the mathematical problem has continued. With the advent of high-speed computers, the investigation of these configurations intensified, especially among physicists, chemists and crystallographers; see, for example, [2, 3, 7, 9, 10, 15, 16, 17, 23, 24, 27, 28, 46, 48].

Further, the discovery of carbon fullerenes[1] (C_{60}, C_{70}, etc.) and their connection to these equilibrium configurations has provided added impetus to these investigations; see [4, 14, 22, 32, 33, 37, 49]. For example, the structure of the stable carbon-60 molecule has 60 atoms on the 60 vertices of the icosa-dodecahedron (soccer ball), which coincides with the cell structure of 32 electrons in equilibrium on the sphere; see Fig. 6. We will see in Section 6 that this is not an accident.

---

[1]Curl, Kroto, and Smalley received the 1996 Nobel Prize in Chemistry for their work on fullerenes

Most of the problems involve finding the configurations of N points on the surface of the unit sphere that maximize or minimize some given quantity (criterion). Of course, we wish to know whether the extremal configurations are unique, modulo rotations and reflections. Also of interest are local maxima or minima; that is, configurations for which very small perturbations reduce or increase (resp.) the criterion. Sometimes (particularly for small values of N), but certainly not always, there is just one local extremum which, therefore, must be the global one.

Many questions arise concerning the placement of points on the sphere "uniformly" with respect to surface area, or the partition of the sphere into regions in a regular way, etc. In general, when we wish to find N points that optimize a certain energy function, the best we can hope for is to find the exact solutions only for small N. However, there may be some rather large values of N when the problem can be solved using certain highly symmetric configurations of points. Even when the explicit solutions cannot be found, it is often still possible to make qualitative statements about the optimal configurations.

The outline of this chapter is as follows. Section 2 begins with some notation and a short survey of various problems of arranging points on the sphere. The approach of estimating the extremal energies via partitioning the sphere into N parts with equal areas with minimal diameters is reviewed in Section 3. In Section 4, we discuss our numerical experiments and the algorithms we implemented. For $N = 2, \ldots, 200$, computational results for the three classical cases $\alpha = 0, \pm 1$ are presented via easily comprehended 2-D plots. Also, the behavior of the extremal energies is investigated. We present a general conjecture on the asymptotic form (as $N \to \infty$) of the extremal energy, supported by numerical evidence. We also investigate the interesting geometric structures and group properties of the extremal points, their cell structure and close relation to the carbon fullerenes.

Finally, we make use of the numerical evidence that the computed extremal points have certain symmetry properties, and that they try to imitate points in a regular hexagonal tiling of the plane. These observations lead us to an explicit formula for generating any number of points, which we call "generalized spiral points", that yield good estimates for the extremal energy. At least for $N \le 12\,000$, these points provide a reasonable solution to a problem of M. Shub and S. Smale [36] arising in complexity theory.

2 Notation, Terminology, and Problems

A set of $N \ge 2$ points $\{x_1, \ldots, x_N\}$ on the unit sphere $S^2 := \{x \in \mathbb{R}^3 : |x| = 1\}$ will be denoted by ω_N. We shall use $|x - y|$ to denote the Euclidean distance in $\mathbb{R}^3$ between two points $x, y \in S^2$. For each real α, the α-*energy* associated with

ω_N is defined by

$$E(\alpha, \omega_N) := \begin{cases} \sum_{1 \le i < j \le N} \log \frac{1}{|x_i - x_j|}, & \text{if } \alpha = 0 \\ \sum_{1 \le i < j \le N} |x_i - x_j|^\alpha, & \text{if } \alpha \ne 0. \end{cases} \quad (2.1)$$

Our concern is with the *extremal energy* for N points on the sphere:

$$\mathcal{E}(\alpha, N) := \begin{cases} \inf_{\omega_N \subset S^2} E(\alpha, \omega_N), & \text{if } \alpha \le 0 \\ \sup_{\omega_N \subset S^2} E(\alpha, \omega_N), & \text{if } \alpha > 0. \end{cases} \quad (2.2)$$

The determination of $\mathcal{E}(\alpha, N)$ is an important and active research area (see the survey paper by Melnyk et al. [27] and the expository paper by Saff and Kuijlaars [31]). Since S^2 is compact, it is clear that for each $N \ge 2$, the extremal α-energy is attained by some point set, which we denote by $\omega_N^{(\alpha)}$. Such equilibrium points are not unique, but we use $\omega_N^{(\alpha)}$ to denote any particular determination of them.

2.1 Various Problems of Arranging Points on the Sphere

Most of the following various mathematical problems of selecting N points on the sphere arise directly from physical applications, in such diverse fields as crystallography, electrostatics, geoscience, medicine, molecular structure, numerical integration, and viral morphology.

1. *Maximum average distance* ($\alpha = 1$). Determining the exact value of $\mathcal{E}(1, N)$, is, except for certain small values of N, a long-standing open problem in discrete geometry, which was initiated by L. Fejes Tóth [19]. This problem is still open for $N > 4$. Berman and Hanes [9] prove a lemma and use it to implement an iterative program, yielding results for N up to 10. Recently, several authors, including Alexander [1], Stolarsky [38, 39], and Beck [6] have made significant contributions; see Stolarsky [39] for history.

2. *The electron problem* ($\alpha = -1$). The determination of $\mathcal{E}(-1, N)$ is called J. J. Thomson's Problem. Apparently, he was the first person who clearly stated the problem; see e.g. Thomson [41]. This problem is also known as the *Coulomb potential* or *electron problem*. Optimizing points are in this case referred to as *Fekete points* in [15]. Each of the N points on the unit sphere can be considered as the position of a unit electric charge, and we are asked to minimize the total energy. Föppl [21], at the suggestion of Hilbert, made a rigorous examination of Thomson's arrangements. He gave answers for $N = 5, 6, 7, 10, 12, 14$. Cohn computed results for $N = 9$ and 11; see L. L. Whyte [47] for history. Partly due to the recent discovery of carbon fullerenes (C_{60}, C_{70}, etc., see [14], [37]), this problem has again attracted the attention of researchers in chemistry, physics, and crystallography. There are hundreds of references to Thomson's Problem and its applications. Here we cite only a few recent ones (of more mathematical content): [2, 3, 7, 10, 15, 16, 17, 23, 24, 27, 28, 46, 48, 49].

3. *Maximize the product of distances* ($\alpha = 0$). For $\alpha = 0$, the minimum energy problem is easily seen to be equivalent to the problem of maximizing the product of the $n(n-1)/2$ distances between pairs of the N points on the sphere. This problem is open for $N > 4$. For N large, G. Wagner [42, 43, 44, 45] obtained upper and lower bounds for the maximal product.

It is easy to see that this problem is equivalent to the problem of minimizing the energy with logarithmic kernel. The logarithmic potential is of interest when considering the thermodynamic limit of many particles on a large sphere; see [8].

Shub and Smale in [34, 35, 36] posed the problem of finding explicit sets of points $\omega_N = \{x_1, \ldots, x_N\} \subset S^2$ that are near optimal in the sense that, for some constant C_0, we have

$$E(0, \omega_N) - \mathcal{E}(0, N) \leq C_0 \log N, \quad \forall\ N \geq 2. \tag{2.3}$$

Such points are of interest since they serve as good starting values for Newton's method; see [36].

Recently, B. Bergersen et al. [8] also considered this problem; they obtained energy values for $2 \leq N \leq 65$. Their results agree with ours except in the case of $N = 46$, where our energy value is lower. So far, we have not encountered any better results than those of our computations. We have an extensive table of values of $2 \leq N \leq 200$; refer to Table 2 in [50].

4. *The packing problem* ($\alpha = -\infty$) It is interesting to note that as $\alpha \to -\infty$, the minimal discrete energy problem tends to the Best Packing Problem on the sphere (also known as Tammes' Problem [40]), which asks for the largest spherical radius of N identical spherical caps that can be packed onto the surface of the unit sphere; or equivalently, to maximize the least distance between any pair of points. G. Fejes Tóth and L. Fejes Tóth [18] refer to this problem as the "problem of inimical dictators" since it can be stated in the following form: "A spherical planet is governed by N mutually inimical dictators. How should they place their residences in order to get as far as possible from one another?" The literature for this problem is enormous, so we refer the interested readers to the article of Coxeter [12] and the books of L. Fejes Tóth [20] and Conway and Sloane [11].

3 Equal-Area Partitions and Extremal Energy Estimates

We proved in [50] following theorem:

Theorem 3.1 *For any $\varepsilon > 0$, there exist a k_0 and N_0 depending on ε, such that for any $N \geq N_0$, we can find an equal-area partition $\mathcal{D} = \{D_i\}_{i=1}^N$ of S^2 with the following properties:*

$$d(D_i) \leq \begin{cases} 2\sqrt{2\pi}\,(1+\varepsilon)/\sqrt{N}, & i = k_0+1, \ldots, N-k_0, \\ 7/\sqrt{N}, & \text{otherwise.} \end{cases}$$

The usefulness of Theorem 3.1 is made clear by the following simple result.

Theorem 3.2 Let $K(r)$ be a lower semi-continuous decreasing function for $0 < r \leq 2$ and suppose

$$\beta(K) := \frac{1}{(4\pi)^2} \iint_{S^2 \times S^2} K(|x-y|) d\sigma(x) d\sigma(y) < \infty. \tag{3.1}$$

If $\mathcal{D} = \{D_i\}_{i=1}^N$ is an equal-area partition of S^2 into N parts, then there exist points $\{\hat{x}_i\}_{i=1}^N$ with $\hat{x}_i \in D_i$, $i = 1, \ldots, N$, such that

$$\sum_{1 \leq i \neq j \leq N} K(|\hat{x}_i - \hat{x}_j|) \leq N^2 \beta(K) - \sum_{i=1}^N K(d(\mathring{D}_i)). \tag{3.2}$$

Remark 3.3 If K is increasing and upper semi-continuous, then (3.2) is true with the inequality sign reversed.

It is easy to verify that for the kernel $K(r) = r^\alpha$, $-2 < \alpha < 2$, $\alpha \neq 0$, we have

$$\frac{1}{4\pi} \int_{S^2} |x-y|^\alpha d\sigma(x) = \frac{2^{\alpha+1}}{2+\alpha}, \quad y \in S^2;$$

hence $\beta(K) = 2^{\alpha+1}/(2+\alpha)$ (cf. (3.1)), and so from Theorems 3.2 and 3.1 we deduce the following.

Corrollary 3.4 Given $-2 < \alpha < 2$, $\alpha \neq 0$, and $\varepsilon > 0$, there exists an $N_0 = N_0(\varepsilon, \alpha)$ such that for any $N \geq N_0$,

$$\mathcal{E}(\alpha, N) \leq \frac{2^\alpha}{2+\alpha} N^2 - \frac{1}{2}(2\sqrt{2\pi})^\alpha (1-\varepsilon) N^{1-\alpha/2}, \quad \text{if } -2 < \alpha < 0,$$

$$\mathcal{E}(\alpha, N) \geq \frac{2^\alpha}{2+\alpha} N^2 - \frac{1}{2}(2\sqrt{2\pi})^\alpha (1+\varepsilon) N^{1-\alpha/2}, \quad \text{if } 0 < \alpha < 2.$$

Theorem 3.5 For $-2 < \alpha < 2$, define B_α by

$$\mathcal{E}(\alpha, N) - \left[-\frac{1}{4} \log\left(\frac{4}{e}\right) N^2 - \frac{1}{4} N \log N \right] = B_\alpha N, \quad \text{if } \alpha = 0, \tag{3.3}$$

$$\mathcal{E}(\alpha, N) - \frac{2^\alpha}{2+\alpha} N^2 = B_\alpha N^{1-\alpha/2}, \quad \text{if } \alpha \neq 0. \tag{3.4}$$

Then we have:

$$-0.1127688 \leq B_0 \leq -0.0234973; \tag{3.5}$$

$$-\frac{5+2\alpha}{4+2\alpha} \leq B_\alpha \leq -\frac{1}{2}\left(2\sqrt{2\pi}\right)^\alpha, \quad \text{if } -2 < \alpha < 0; \tag{3.6}$$

$$-\frac{1}{2}\left(2\sqrt{2\pi}\right)^\alpha \leq B_\alpha < 0, \quad \text{if } 0 < \alpha < 2. \tag{3.7}$$

Therefore, the magnitude of $\mathcal{E}(\alpha, N)$ grows with an order of N^2.

4 Numerical Experiments, Asymptotics of Extremal Energy

To determine extremal α-energies numerically we begin by mapping the sphere onto the extended complex plane via stereographic projection. This permits the use of methods of unconstrained optimization.

We did extensive numerical experiments over a two-year period. We began with the case $\alpha = 0$. As a first step, the Steepest Descent method was examined. To expedite this algorithm for our particular case, we extracted an explicit formula for the descent step; so as to avoid the expensive part of this algorithm, i.e., the line minimization required to calculate the step size in each iteration. However, we found convergence was slow near the extremal points when high precision was desired. To accelerate the convergence, we mimicked Newton's method, using first-order linear approximation in order to avoid calculating the inverse of a Jacobian matrix. When we combined the two algorithms, using the Steepest Descent for rough calculation and then switching to the Newton-type algorithm, we obtained a robust and fast algorithm. Unfortunately, this technique was successful only for the case $\alpha = 0$.

To obtain a general method to optimize the energy, we investigated the Downhill Simplex method, the Conjugate Gradient method and the Quasi-Newton method (or Variable Metric method). We refer the reader to [29, 30] for more information about these algorithms. We found that for the energy problem, the Quasi-Newton method worked best. We implemented this method so that it would calculate the extremal α-energy for any α. A few observations are in order from our calculations.

First, when N is large, say larger than 60, the number of equilibrium metastable states increases dramatically. Furthermore, to make things even more difficult, the energy surface is very flat, and all the local extrema have energies which are very close numerically. To give a rough idea, we plot in Fig. 1 the energy value difference, $E(0, \omega_N) - \mathcal{E}(0, N)$ for all the equilibrium states that we found for $\alpha = 0, 4 \leq N \leq 122$, where $E(0, \omega_N)$ is the energy for a particular local extremal point set ω_N. The horizontal axis represents values of N. Each point represents a local extremum. As one can see from the scale, even as N gets larger, all the local extremal points have very close energies. Therefore, it is increasingly difficult to find a global extremum as N becomes large.

To maximize the possibility that the extremal points found were actual global extremal points, we tried different iteration schemes and at least 1000 different random start positions. We also examined the geometric and group structure of the extremal points, and used the appropriate asymptotic approximation formula (see (3.7)) to inspect any suspicious energy values obtained. We also ran cross-checks in the following way. For each N, we compared the symmetry groups for the extremal points for $\alpha = 0, \pm 1$. If these groups were different, for example if the symmetry groups for $\omega_N^{(0)}$ and $\omega_N^{(-1)}$ were not the same, we would substitute $\omega_N^{(0)}$ as the starting position to find $\omega_N^{(-1)}$, and then compare the final energy

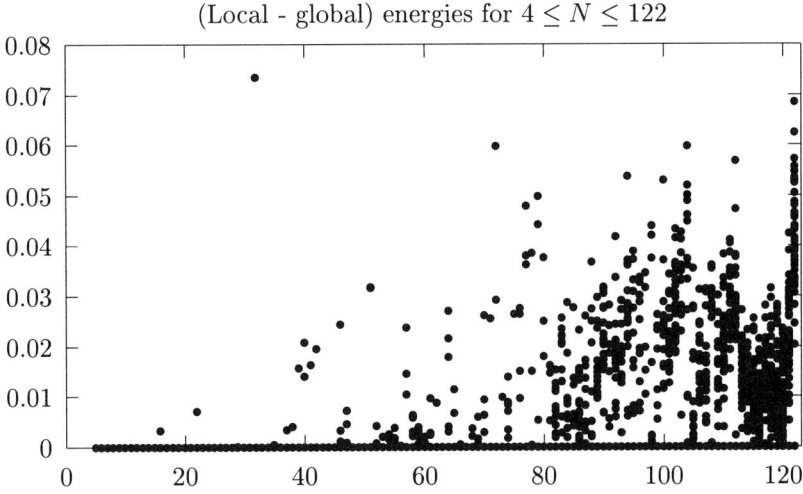

FIG. 1. Distributions for the local extremal energies

values and group properties. A similar procedure was employed in all possible cases. Second, we found that the extremal points for different values of α are usually very close; typically, they give even better energy values than many of the corresponding local extrema.

Based on the numerical results, we obtained the following numerical asymptotic approximations $f(\alpha, N)$ to the actual values of $\mathcal{E}(\alpha, N)$ for $\alpha = 0, \pm 1$ and $N \leq 200$.

$$f(-1, N) = \frac{N^2}{2} - 0.55230 N^{3/2} + 0.0689 N^{1/2}, \tag{4.1}$$

$$f(0, N) = -\frac{1}{4} \log\left(\frac{4}{e}\right) N^2 - \frac{1}{4} N \log N - 0.026422 N + 0.13822, \tag{4.2}$$

$$f(1, N) = \frac{2}{3} N^2 - 0.40096 N^{1/2} - 0.188 N^{-1/2}. \tag{4.3}$$

To evaluate our approximation formulas (4.1), (4.2), (4.3), we plot the difference $\mathcal{E}(\alpha, N) - f(\alpha, N)$ for $\alpha = 0, \pm 1$ and $4 \leq N \leq 200$ in Figures 2, 3, and 4. We used the gnuplot program with the impulse plot-style, which simply draws vertical lines from the N-axis to the points being plotted. It is important to note the significance of these plots. As we know from the inequalities (3.5), (3.6), the growth of $\mathcal{E}(\alpha, N)$ is of order N^2, and Figures 2–4 suggest that the differences $\mathcal{E}(\alpha, N) - f(\alpha, N)$ are bounded by a very small constant times $N^{-\alpha/2}$.

5 Geometry of Extremal Points and Carbon Fullerenes

5.1 Carbon Fullerenes

At high school, we were pedagogically taught that there were only two forms of pure carbon, graphite and diamond. So we can all imagine the shock when

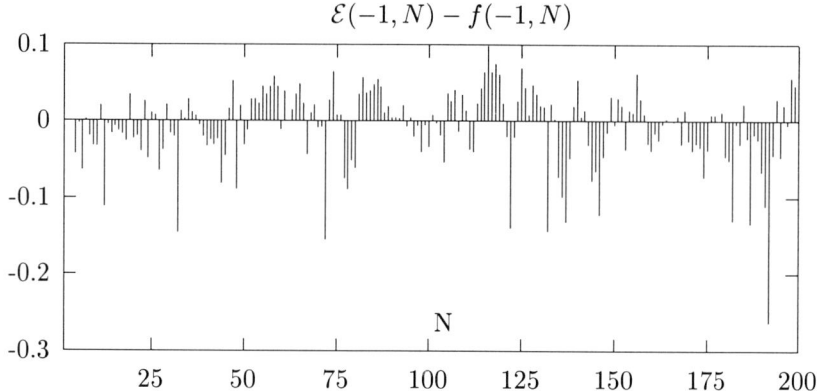

Fig. 2. Error in approximating extremal Coulomb energy

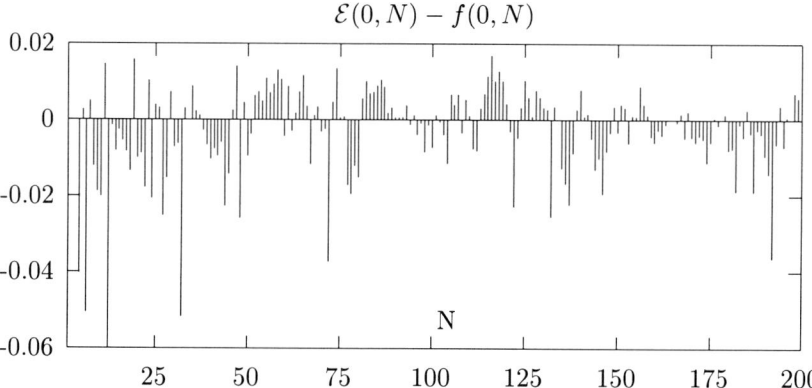

Fig. 3. Error in approximating extremal logarithmic energy

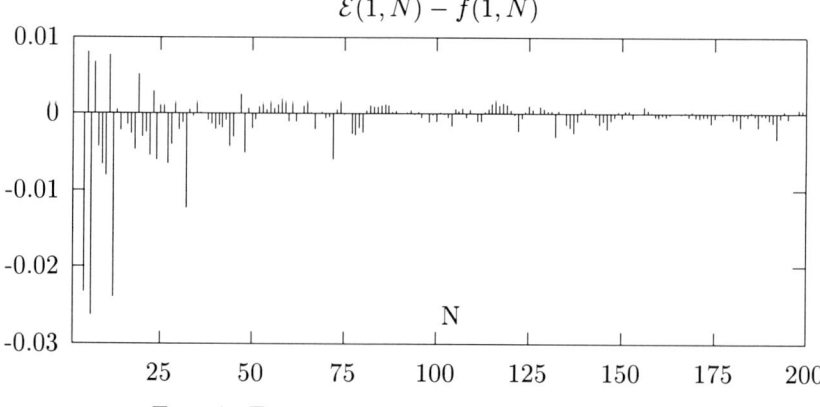

Fig. 4. Error in approximating extremal sums

Richard E. Smalley, Harry Kroto, and their associates announced in 1985 (see [26]) that they discovered a new form of pure carbon. The excitement came in 1990, when a convenient way of making carbon C_{60} was discovered and it turned out that this molecule is a hollow cluster of 60 carbon atoms shaped like a soccer ball; see [14] and [25]. This new carbon molecule is also called Buckminsterfullerene or buckyball — named for the American architect R. Buckminster Fuller, whose geodesic domes had a similar structure.

The discovery set off a research boom on the structure and electronic properties of the fullerenes. Joined in this "gold rush" are chemists, physicists, materials scientists, and even mathematicians.

5.2 Geometry of the Extremal Points

As one can see from Table 2 in [50], most of the extremal point sets have symmetry properties. We now look at these points from another perspective. Let $\omega_N = \{x_1, x_2, \ldots, x_N\}$ be a set of N points on S^2. Define

$$D(x_i) := D_i := \{x \in S^2 : |x - x_i| \leq |x - x_j|, \ \forall j\},$$

which is the *Dirichlet cell* of x_i. In other words, D_i is the region of the sphere that consists of points that are closer to x_i than to any other points of ω_N. It is easy to prove that each D_i is a spherical polygon, i.e., the boundary of D_i consists of finitely many pieces of great circle arcs. We say D_i is a *spherical r-gon* if the boundary of D_i consists of r great circle arcs. Numerical evidence for N up to 200 strongly suggests the following.

Conjecture 5.1 *If $N \geq 6$, then the Dirichlet cells of $\omega_N^{(\alpha)}$ consist only of spherical 4-gons, 5-gons, and 6-gons, i.e., quadrilaterals, pentagons, and hexagons.*

If the Dirichlet cells for $\omega_N^{(\alpha)}$ consist only of pentagons and hexagons, it is an easy consequence of Euler's identity that there are exactly 12 pentagons, provided that exactly 3 edges emanate from each vertex. The total number of vertices is $2(N - 2)$.

Our computations suggest that the 12 points whose Dirichlet cells are pentagons tend to distribute themselves as far apart from each other as possible. While they usually do not form the vertices of an icosahedron, the arrangement is quite close to that. Figure 5 depicts 32 and 122 electrons in equilibrium on the surface of the sphere and their Dirichlet cells for the case $\alpha = -1$. The pentagons in the case $N = 32$ yield nearly an icosahedral arrangement, while the pentagons in the case $N = 122$ are numerically indistinguishable from an exact icosahedral arrangement.

As one can see, the cell structure of 32-electrons resembles the C_{60} fullerenes. Further the cell structure of 122-electrons resembles the postulated C_{240}. In fact, they are very close. Zhang et al. [49] combined this idea in attempting to determine numerically the structures for C_{20} up to C_{70}.

If we don't draw in the sphere and the electrons, but simply connect the junctions (vertices) of the cell structure with straight segments and put an atom

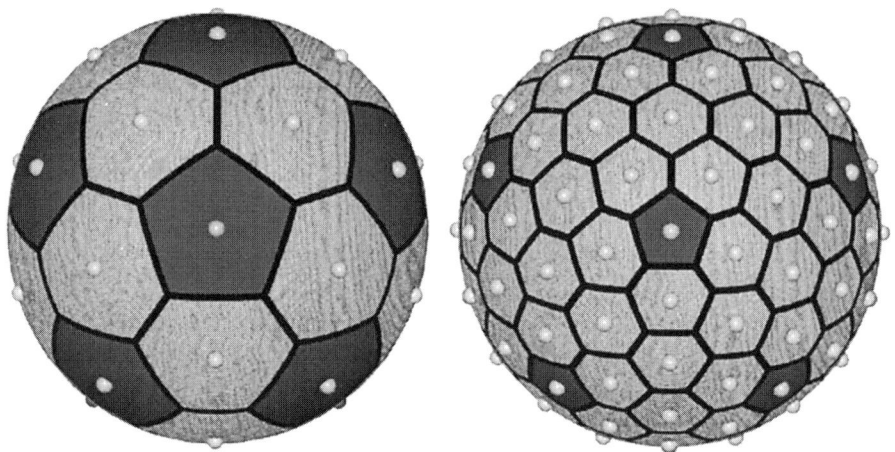

FIG. 5. 32 electrons and 122 electrons in equilibrium on S^2

at each junction point of the cell, we obtain figures that resemble the carbon fullerene structure. We call this structure the *dual structure* of the extremal energy points. It is easy to see that, given N extremal points, and assuming the dual has only hexagonal and pentagonal cells, then we have 12 pentagons and $N-12$ hexagons and $2(N-2)$ junction points (atoms). For example, 32-electrons produce a candidate for C_{60}, 37-electrons produce a candidate for C_{70}, 72-electrons yield a candidate for C_{140} and 122-electrons for C_{240}. From our calculations, we also know that when the number of electrons is 32, 72, 122, 132, 192, 272, 282, the extremal points have icosahedral symmetry. So we predict that we could also have stable C_{140}, C_{240}, C_{260}, C_{380}, C_{540} and C_{560}. Here we only sample two cases 'C_{60}' and 'C_{240}'; see Fig. 6.

Of course we know that, except C_{60}, the other fullerenes will not form a perfect spherical shape; rather they shape like an oblong football. But the cell structures have 12 pentagons and the remaining cells as hexagons. Thus we conjecture that the dual structure of the extremal electron problem will at least have the same combinatorial network structure as that of corresponding carbon fullerenes.

6 A Class of Explicit Points that Equally Distribute over the Surface of the Sphere

The problem of generating N points of the sphere that are equally distributed has many different roots, from such diverse fields as various virus morphology, modeling the structure of carbon fullerenes, geoscience, medicine, and numerical integration of functions on the sphere. Also, we are pleased to say that these general spiral points were used as starting points for various iteration schemes. They are especially powerful when large number of points are desired.

From numerical experiments, it appears that the equilibrium points attempt

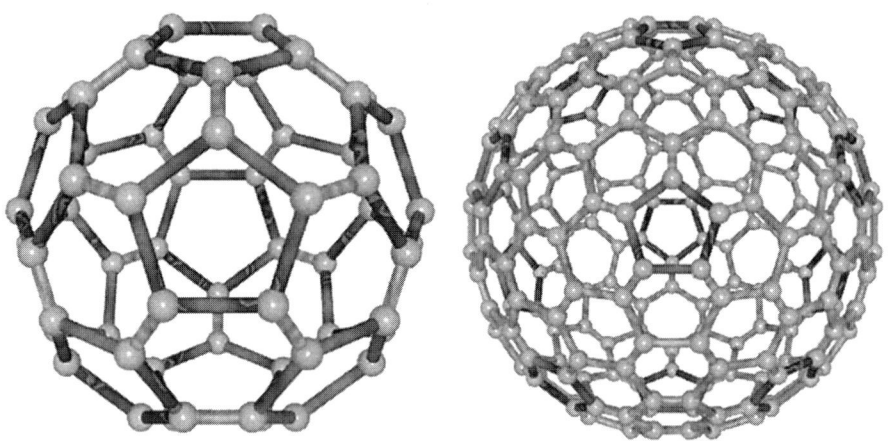

FIG. 6. Dual structure of 32 and 122 electrons in equilibrium on S^2

to distribute themselves over a nearly regular spherical hexagonal net. We devised a simple scheme for imitating this behavior for any given N. Compared with various schemes in literature, the generalized spiral points are superior in the setting of extremal energy problems.

6.1 Construction of the Generalized Spiral Points

To describe these points we use the spherical coordinates (θ, ϕ), $0 \leq \theta \leq \pi$, $0 \leq \phi \leq 2\pi$. Let

$$h_k := -1 + 2(k-1)/(N-1), \quad 1 \leq k \leq N; \quad \theta_k := \arccos(h_k); \tag{6.1}$$

$$\phi_1 := \phi_N := 0, \quad \phi_k := \left(\phi_{k-1} + \frac{C}{\sqrt{N}} \frac{1}{\sqrt{1-h_k^2}}\right) (\text{mod } 2\pi), \quad 2 \leq k \leq N-1, \tag{6.2}$$

where the constant $C = \sqrt{8\pi/\sqrt{3}}$ is chosen so that successive points will have approximately the same (Euclidean) distance apart on S^2. The point set $\hat{\omega}_N = \{(\phi_k, \theta_k)\}_{k=1}^N$ is called a *generalized spiral* on S^2.

However, based on numerical calculation, we found that choosing $C = 3.6$ in (6.2) produces better energy approximation. This can viewed as accommodation to the fact that distances shrink when a hexagon is projected onto the surface of the sphere. For this choice, we plot the difference $E(0, \hat{\omega}_N) - f(0, N)$ for $N \leq 12,000$, and we obtain Fig. 7. This figure shows that the generalized spiral points have, for large N, energy that agrees with (3.5) to within $O(N)$. In fact, although $\hat{\omega}_N$ does not appear to solve the Shub and Smale Problem (cf. (2.3)), numerically it gives

$$E(0, \hat{\omega}_N) - f(0, N) \leq (5/2) \log N, \quad \text{for } 2 \leq N \leq 12\,000.$$

Furthermore, from the estimate (3.5) and the computed values of $E(0, \hat{\omega}_N)$, we find
$$E(0, \hat{\omega}_N) - \mathcal{E}(0, N) \le 114 \log N, \quad \text{for } 2 \le N \le 12\,000. \tag{6.3}$$

For $-2 < \alpha < 2, \alpha \ne 0$, computations indicate that these same spiral points have α-energy that agrees with (4.1) to within $O(N^{1-\alpha/2})$. It is really astonishing that such a simple scheme of system points will approximate the extremal energy problem so well.

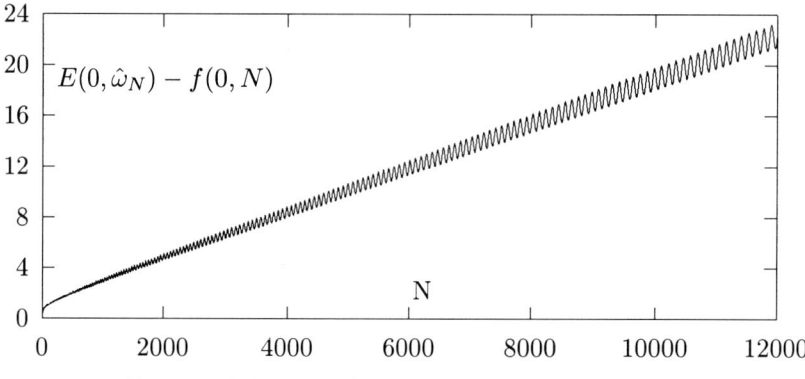

FIG. 7. Behavior of energy for generalized spiral

6.2 How Good are the Generalized Spiral Points?

Due to the unexpected success of the generalized spiral points in the setting of extremal energy problems, we considered the issue of how good the generalized spiral points are in general. Are they evenly distributed? We can do several things to get different perspectives.

First, we plotted the generalized spiral points for various values of N. Certainly we want points that distribute uniformly on the sphere (with respect to area). As one can see from Fig. 8, the generalized spiral points appear to distribute themselves very regularly over the surface of the sphere.

We have to acknowledge that the spirals tend to have mild concentration near the two poles; see the pole view of the right figure in Fig. 8. There are a number of ways to remedy that. For example, one way is to adjust the formula for the first few and the last few points. Another way is, instead of putting 1 point at each level, we can put 5 or 6 points evenly spaced at each horizontal level. The resulting formula is a bit more complicated, and that will work for a subsequence of points. In general, that will eliminate the problem of concentration near the poles and yield slightly better results. In subsection 6.3, we investigate a variant of the generalized spiral points that will require some computation, but generally yields better results.

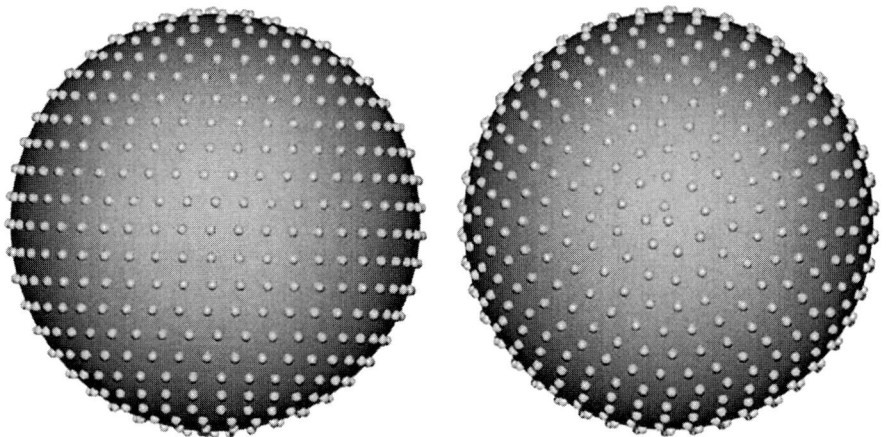

FIG. 8. Side and pole view of generalized spiral points for $N = 702$

Second, we compared them with another popular, reasonable way of generating "good" points on the sphere called "Dissection of Icosahedron" [5], which can generate a subsequence of $\{10 \cdot 4^n + 2\}$ points, for $n = 0, 1, 2, 3, \ldots$. We compared the energies of these two different point sets and we tabulated the results, taking the logarithmic potential case ($\alpha = 0$) as an example.

N	Icosa-Dissection	Spiral	Difference	Percentage of Difference
42	-210.522630	-210.235921	0.286709	-0.136189
162	-2742.027879	-2744.002145	-1.97427	0.0720002
642	-40813.642692	-40856.875926	-43.2332	0.105928
2562	-638267.760101	-638983.072529	-715.312	0.112071
10242	-10142813.267666	-10154326.307054	-11513.0	0.113509

Table 1 Energy differences between Icosa-Dissection and the spiral points

It is clear from Table 1 that, except for $N = 42$, our generalized spiral points have much smaller energy, if one considers the fact that the energy function surface is very flat. We could also look at Fig. 7, and observe that for $N = 10\,242$, the energy difference between the generalized spiral points and the projected extremal value is about 20. On the other hand, the logarithmic energy of the points generated by the Icosa-Dissection procedure is 11 513 higher than that of the generalized spiral points.

Third, J. Cui and W. Freeden [13] developed a concept of generalized discrepancy, which involves pseudodifferential operators to give a criterion for equidistributed pointsets on the sphere.

Definition 6.1 *Given a point set $\omega_N = \{x_1, \ldots, x_N\}$, define the discrepancy*

$\mathcal{D}(\omega_N)$ of ω_N as:

$$\mathcal{D}(\omega_N) := \frac{1}{2\sqrt{\pi N}} \left[\sum_{i=1}^{N} \sum_{j=1}^{N} \left(1 - 2\ln\left(1 + \sqrt{\frac{1 - x_i \cdot x_j}{2}}\right) \right) \right]^{1/2} \qquad (6.4)$$

It was argued in [13] that for given N, the smaller the value of $\mathcal{D}(\omega_N)$ is, the more evenly distributed the points are. Of the 5-point systems investigated in [13], the superior one is the Hammersley system. Therefore, we compare Hammersley points with the general spiral points and list the results in Table 2.

N	Hammersley points	General spiral points
5	0.0813	0.0886
23	0.0262	0.0257
45	0.0157	0.0152
85	0.0097	0.0093
125	0.0072	0.0069
165	0.0060	0.0056
218	0.0048	0.0045
250	0.0043	0.0041
296	0.0038	0.0036
350	0.0034	0.0032
394	0.0031	0.0029
440	0.0028	0.0027

Table 2 Discrepancies of Hammersley points and general spiral points

In fact, the general spiral points algorithm is very stable. If we vary the constant C in (6.2), from 3.2 to 4, the point system generated is still better than those of Hammersley points relative to the criteria of discrepancy.

6.3 A Variant of the Generalized Spiral Points

A natural question is: can we do better than the generalized spiral points? The answer is yes, if we are willing to do some short calculation at the beginning.

In this subsection, we introduce another way of getting N points equally spaced on the sphere. Since the way to generate N points evenly spaced on the sphere is quite useful in many fields, we think it appropriate simply to include a short code segment here. The following code, given appropriate arguments, will generate the three-dimensional Euclidean coordinates and store them in the array.

```
typedef struct point3 { double x; double y; double z; } point3;

/* N    -- # of points desired
   m    -- how many points on the same level.
   pole -- 0, 1, 2 poles included in the points
   adj  -- inital angle adjustment, a number between 2.37 and 4.15
```

```
      *p       -- pointer to the storage for N points of struct point3
*/
double spiral(int N, int m, int pole, double adj, point3 *p)
{ int i, j, k=0; int *M;
  double s, theta, r, x, y, t;

  n=(N-pole)/m;   /* integer division */
  M =(int *)malloc( (n+1) * sizeof( int ) );
  M[0]=M[n+1]=0;
  if( pole == 2) { M[0]=M[n+1]=1;}
  else if { (pole == 1) M[n+1]=0;}
  for(i=0; i<n; i++) { M[i+1]=m; }
    s=(double)M[0]*0.5;
    if( pole == 2 ) {
      p[k].x=p[k].y=0.0; p[k].z=1.0; k++;
  }
  for(i=1; i<n+1; i++)
  { s+=0.5*M[i]+0.5*M[i-1];
    r=sqrt( (N-s-0.5)/(s-0.5) );
    theta=2.0*(i-1)*M_PI/(m*adj);
    t=r*r;
    for(j=0; j<M[i]; j++) {
      theta+=2.0*M_PI/m;
      p[k].x=2.0*r*cos( theta)/(1.0+t);
      p[k].y=2.0*r*sin( theta)/(1.0+t);
      p[k].z=(t-1.0)/(t+1.0);
      k++;
    }
  }
  if( pole ) { p[k].x=p[k].y=0.0; p[k].z=-1.0; }
}
```

One thing worth mentioning is that this procedure is quite sensitive to the initial angle adjustment (*adj* in the code above). If you have a criterion, say, minimizing the energy, what can be done is to incorporate the criterion in conjunction with a very simple line minimization for *adj* between 2.375 and 4.125. Typically, after the line minimization, these points yield better results over the generalized spiral points.

Let's plot the 702 points obtained via the above code with $m = 1, pole = 2$; the result is Fig. 9. The left figure is the side view, and the right one is the pole view (bottom view). We can still see that the points tend to have mild concentration near the two poles.

In this case, it is also very easy to remedy that. Simply take $m = 5, pole = 2$; the points are better than those if we take $m = 1, pole = 2$, both in terms of even distribution near the poles and low energy; see Fig. 10.

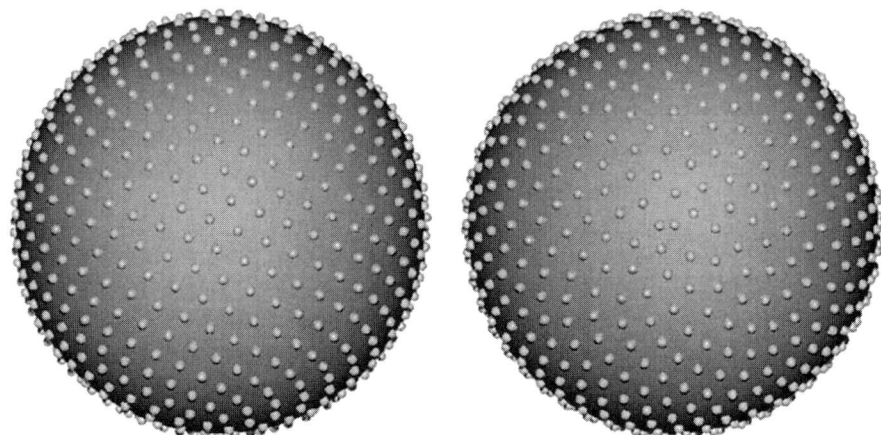

FIG. 9. Side and pole view for $N = 702$ and $m = 1, pole = 2$

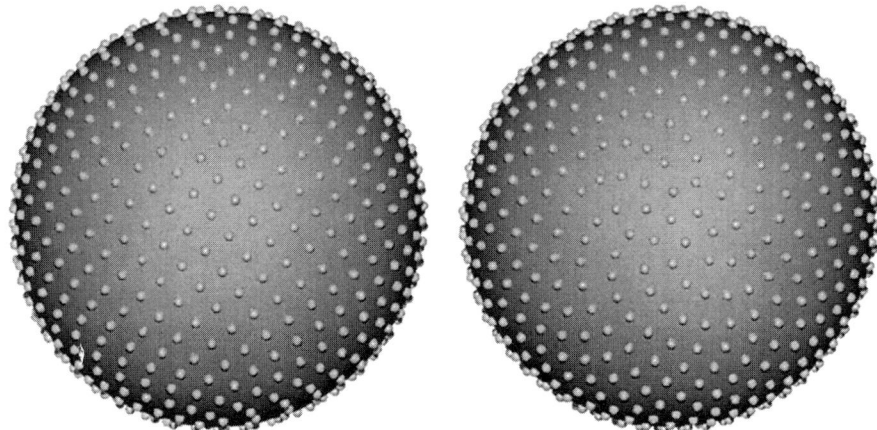

FIG. 10. Side and pole view for $N = 702$ and $m = 5, pole = 2$

Interested readers are encouraged to visit website:
http://www.math.nwu.edu/ zhou/
where they can find more detailed discussion. Also, a very extensive graphics library, including different perspectives of all the points for N from 2 to 200, can be found.

Acknowledgements

The author would like to acknowledge the tremendous effort by Professor Jerome in making possible such a wonderful workshop. I am grateful to him for the opportunity to participate in the workshop. I am also grateful for his extensive suggestions concerning this article. I also want to thank E.A. Rakhmanov and E.B. Saff for their mentoring and friendship.

Bibliography

1. Alexander, R. (1972). On the sum of distances between n points on a sphere, *Acta Math. Acad. Sci. Hungar.*, **23**, 443–448.
2. Alexander, R. and Stolarsky, K. B. (1974). Extremal problems of distance geometry related to energy integrals, *Trans. Amer. Math. Soc.*, **193**, 1–31.
3. Ashby, N. and Brittin, W. E. (1986). *Amer. J. Phys.*, **54**, 776–777.
4. Babić, D., Klein, D. J., and Sah, C. H. (1993). Symmetry of fullerenes, *Chem. Phys. Letters*, **211**, 235–241.
5. Baumgardner, J. R. and Frederickson, P. O. (1985). Icosahedral discretization of the two-sphere, *SIAM J. Numer. Anal.*, **22**, 1107–1115.
6. Beck, J. (1984). Some upper bounds in the theory of irregularities of distribution, *Acta Arithm.*, **43**, 115–130.
7. Berezin, A. A. (1986). Asymptotics of the maximum number of repulsive particles on a spherical surface, *J. Math. Phys.*, **27**, 1533–1536.
8. Bergersen, B., Boal, D., and Palffy-Muhoray, P. (1994). Equilibrium configurations of particles on a sphere: the case of logarithmic interactions, *J. Phys. A: Math. Gen.*, **27**, 2579–2586.
9. Berman, J. and Hanes, K. (1977). Optimizing the arrangement of points on the unit sphere, *Math. Comput.*, **31**, 1006–1008.
10. Calkin, M. G., Kiang, D., and Tindall, D. A. (1987). Minimum-energy charge configurations, *Am. J. Phys.*, **55 No.2**, 157–158.
11. Conway, J. H. and Sloane, N. J. A. (1993). *Sphere Packings, Lattices and Groups*, Springer-Verlag, New York, Berlin, Heidelberg, 2nd edition.
12. Coxeter, H. S. M. (1956). The problem of packing a number of equal non-overlapping circles on a sphere, *Trans. NY Acad. Sci.*, **10**, 117–120, MR 18 #356.
13. Cui, J. and Freeden, W. (1997). Equidistribution on the sphere, *SIAM J. Sci. Comput.*, **18**, 595–609.
14. Curl, R. F. and Smalley, R. E. (Oct, 1991). Fullerenes, *Scientific American*, 54–63.
15. Dahlberg, B. E. J. (1978). On the distribution of fekete points, *Duke Math. J.*, **45**, 537–542.
16. Edmundson, J. R. (1992). The distribution of point charges on the surface of a sphere, *Acta Cryst.*, **A48**, 60–69.
17. Erber, T. and Hockney, G. M. (1991). Equilibrium configurations of N equal charges on a sphere (letter to the editor), *J. Phys. A: Math. Gen.*, **24**, L1369–L1377.
18. Fejes Tóth, G. and Fejes Tóth, L. (1980). Dictators on a planet, *Studia Sci.*

Math. Hung, **15**, 313–316.
19. Fejes Tóth, L. (1956). On the sum of distances determined by a point set, *Acta Math. Acad. Sci. Hungar.*, **7**, 397–401, MR 21 #5937.
20. Fejes Tóth, L. (1964). *Regular Figures*, Pergamon-Macmillan, New York.
21. Föppl, L. (1912). Stabile Anordnungen von Elektronen im Atom, *J. Reine Angew. Math.*, **141**, 251–301.
22. Fowler, P. W. (1986). How unusual is C_{60}? Magic numbers for carbon clusters, *Chem. Phys. Letters*, **131**, 444–450.
23. Frickel, R. H. and Bronk, B. V. (1988). Symmetries of configurations of charges on a sphere, *Can. J. Chem.*, **66**, 2161–2165.
24. Glasser, L. and Every, A. G. (1991). Energies and spacings of point charges on a sphere, *J. Phys. A: Math. Gen.*, **25**, 2473–2482.
25. Kroto, H. (1992). *Angew. Chem. Int. Ed. Engl.*, **31**, 311.
26. Kroto, H., Heath, J., O'Brien, S., Curl, R., and Smalley, R. (1985). *Nature*, **318**, 162.
27. Melnyk, T. W., Knop, O., and Smith, W. R. (1977). Extremal arrangements of points and unit charges on a sphere: equilibrium configurations revisited, *Can. J. Chem.*, **55**, 1745–1761.
28. Munera, H. A. (1986). Properties of discrete electrostatic systems, *Nature*, **320**, 597–600.
29. Polak, E. (1971). *Computational Methods in Optimization*, Academic Press, New York.
30. Press, W. H., Teukolsky, S. A., Vetterling, W. T., and Flannery, B. P. (1971). *Numerical Recipes in C: The Art of Scientific Computing*, Cambridge University Press, Cambridge, New York, second edition.
31. Saff, E. B. and Kuijlaars, A. (1997). Distributing many points on a sphere, *Math. Intelligencer*, **19**, 5–11.
32. Sah, C.-H. (1993). Combinatorial construction of fullerene structures, *Croatica Chemica Acta*, **66**, 1–12.
33. Sah, C.-H. (1994). A generalized leapfrog for fullerene structures, *Fullerene Science & Technology*, **2**, 445–458.
34. Shub, M. and Smale, S. (1993). Complexity of Bezout's theorem I: geometric aspects, *J. Amer. Math. Soc.*, **6**, 459–501.
35. Shub, M. and Smale, S. (1993). Complexity of Bezout's theorem II: volumes and probabilities, *Progr. Math.*, **109**, 267–285.
36. Shub, M. and Smale, S. (1993). Complexity of Bezout's theorem III: condition number and packing, *J. Complexity*, **9**, 4–14.
37. Smalley, R. E. (1991). Great balls of carbon: the story of buckminsterfullerene, *The Sciences*, **31**, (2), 22–28.
38. Stolarsky, K. B. (1972). Sums of distances between points on a sphere, *Proc. Amer. Math. Soc.*, **35**, 547–549.

39. Stolarsky, K. B. (1973). Sums of distances between points on a sphere. II, *Proc. Amer. Math. Soc.*, **41**, 575–582.
40. Tammes, P. M. L. (1930). On the origin of number and arrangement of the places of exit on the surface of pollen-grains, *Recueil des Travaux Botaniques Néerlandais* (Nederlandsche Botanische Vereeniging), **27**, 1–84.
41. Thomson, J. J. (1921). *Philos. Mag.*, **41**, 510.
42. Wagner, G. On a new method for constructing good point sets on spheres (preprint).
43. Wagner, G. (1989). On the product of distances to a point set on the sphere, *J. Austral. Math. Soc. (Series A)*, **47**, 466–482.
44. Wagner, G. (1990). On the means of distances on the surface of a sphere (lower bounds), *Pacific J. Math.*, **144**, 389–398.
45. Wagner, G. (1992). On the means of distances on the surface of a sphere, II (upper bounds), *Pacific J. Math.*, **153**, 381–396.
46. Weinrach, J. B., Carter, K. L., Bennett, D. W., and McDowell, H. K. (1990). Point charge approximations to a spherical charge distribution, *J. Chem. Educ.*, **67**, 995–999.
47. Whyte, L. L. (1952). Unique arrangements of points on a sphere, *Amer. Math. Monthly*, **59**, 602–611.
48. Wille, L. T. (1986). Searching potential energy surfaces by simulated annealing, *Nature*, **324**, 46–48.
49. Zhang, B. L., Wang, C. Z., M., H. K., Xu, C. H., and Chan, C. T. (1992). The geometry of small fullerene cages: C_{20} to C_{70}, *J. Chem. Phys.*, **97**, 5007–5011.
50. Zhou, Y. (1995). *Arrangement of points on the sphere*, Ph.D. thesis, University of South Florida.

4

Some Reduced-Dimension Models Based on Numerical Methods

Todd F. Dupont and A. E. Hosoi
University of Chicago

Abstract

In fluid flow problems, understanding frequently is attained by studying very simplified models. This paper presents a systematic way of generating such models; the techniques involved are inspired by numerical analysis. First we illustrate the process by constructing a simple model of Rayleigh–Bénard convection which gives a remarkably good estimate of the critical Rayleigh number and the corresponding roll size. We also examine a model which helps explain the formation of shocks in creaming or sedimenting colloids. Finally we discuss two models of droplet break-off which are in excellent agreement with experiment and have played a role in the study of cascading instabilities which can occur near snap-off.

1 Introduction

Simplified models of complicated physical systems can be useful in gaining insights into their behavior both analytically and numerically. Here we look at several cases in which one can borrow techniques from numerical analysis to produce such models. In each case we start from one set of partial differential equations and produce a new set that is defined over a space of smaller dimension; hence the use of the term reduced-dimension in the title.

First we look at Rayleigh–Bénard convection in which a liquid in a filled, closed container is heated on the bottom and cooled on the top, and convection rolls develop. We start from a system that is based on a Boussinesq approximation to Stokes flow and is defined in two space dimensions plus time. We produce two different approximate systems to illustrate the techniques involved. In the process we see that while one of these reduced-dimension systems behaves very much like the physical system, the other has some strange behavior. We give some numerical results which show the development of plumes of hot liquid rising to the top of the cell. It is well known that if the temperature difference between the top and bottom is sufficiently small, convection does not occur. A linear stability analysis of the reduced-dimension system is given which predicts critical values for the physical parameters at which rolls start to form. The simplified system is based on the use of a Galerkin method in which only a small amount of flexibility is included in the vertical direction. To indicate the robustness of this approach, we prove an error bound for the Galerkin method in this context that shows the reduced system yields a faithful approximation of the original if enough flexibility is included in the function spaces used.

As a second example we look at a very simple model of what happens when colloidal particles sediment in a liquid. For over a hundred years it has been known that such systems tend to form bands of uniform concentration, and this simple model helps us understand the reason for these bands. The solutions of the system also give a description of the process that brings the bands into being. We first reduce a two space dimensional version of this problem to a one space dimensional version again using a Galerkin technique, and then we show some computational results. Several related techniques, based on numerical methods other than the Galerkin method, are briefly mentioned. Finally, we show how one can reduce a problem that involves three space dimensions in an essential way to one that has only one space dimension.

As a final example we look at axisymmetric free surface flows, at the level of Navier–Stokes with surface tension. Again we use a Galerkin reduction technique as our primary tool. The simplest approximate system that arises in this way of looking at the problem is a set of equations called the Cosserat equations which have been known for many years. It is however rather recent that a strong connection between these equations and the Navier–Stokes equation was established. One gets a family of approximations this way that have a range of flexibilities. We compare the equations derived with the model of Eggers and Dupont for droplet break off.

Our approach is much more numerical in flavor than many commonly used techniques such as asymptotics – i.e. put in enough "grid points" to get reasonable answers. We find these numerically inspired models particularly appealing for several reasons. First, they appear to be enlightening in the physical systems we investigated as described below. Second, we can easily prove convergence theorems using similar arguments to those used in demonstrating convergence in the numerical methods; this is not true in general for simplified models and many asymptotic methods will diverge from the true solution if too many terms are included.

2 Rayleigh–Bénard Convection

To illustrate the application of the dimension reduction techniques we are interested in, we take a simple, but nontrivial, problem of the motion of a liquid heated uniformly from below. This is a well-studied problem and we found the discussions in [3] and [12] to be quite useful.

Take $\Omega = (0, H) \times (0, L)$:

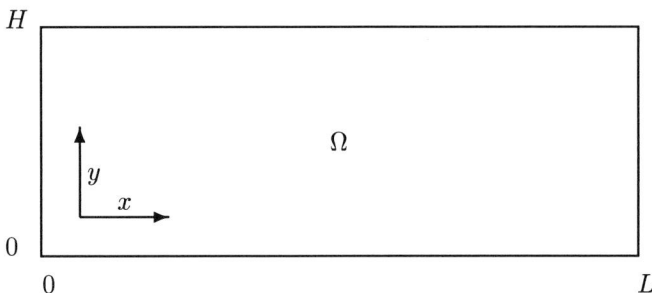

Suppose that the domain, Ω, is filled with a liquid and that we maintain the top and bottom at two different uniform temperatures. The existence of the temperature differential may cause flow and the flow will modify the temperature field in the domain. We suppose that the density of the liquid is almost constant and that inertial forces are very small relative to viscous forces. Under these conditions it is reasonable to assume that the velocity is divergence-free and satisfies the Stokes equation with a forcing term for buoyancy; this is a so-called Boussinesq approximation. The temperature will be taken to satisfy an advection-diffusion equation; we will neglect variations of heat capacity and conductivity and we also neglect the viscous generation of heat.

Adopt the following notation:

$\mathbf{u} = \begin{pmatrix} u_1 \\ u_2 \end{pmatrix}$ denotes the velocity,

T is the temperature,
p is pressure,
ρ is the density,
ν is viscosity,
β is the coefficient of thermal expansion,
D is the thermal diffusivity,
$\mathbf{g}$ is the acceleration of gravity, $\mathbf{g} = g \begin{pmatrix} 0 \\ -1 \end{pmatrix}$.

Then the system we want to study is given by the following set of partial differential equations:

$$\nabla(p/\rho + gy) + \nu \nabla \cdot \nabla \mathbf{u} - \mathbf{g}\beta T = 0, \qquad (2.1)$$
$$\partial_t T + \mathbf{u} \cdot \nabla T - D \nabla \cdot \nabla T = 0, \qquad (2.2)$$
$$\nabla \cdot \mathbf{u} = 0. \qquad (2.3)$$

The velocity is assumed to vanish on the boundary of the domain and the temperature is taken to match $-(\Delta T/H)y$ on the boundary. Because it has the potential for being confusing, we point out explicitly that ρ, the unperturbed density, is assumed constant in this model, even though density variations are what drive the velocity.

As a first observation about this system, note that the temperature satisfies a maximum principle; at no point is its value greater than the larger of its maximum initial value and the maximum on the boundary of the domain. An analogous statement holds for the minimum. Hence if $T(x, y, 0)$ is bounded, then T is bounded for all time. It follows then that the $H_0^1(\Omega)$ norm of **u** is bounded for all time. To see that this is so, just take the inner product of (2.1) with **u** and integrate over Ω; the gradient term integrates to zero since **u** is divergence-free.

Before describing the reduced-dimension model we discuss briefly a Galerkin approximation of the above system, because this type of approximation leads directly to the reduced system. Let $\mathcal{V}$ be a divergence-free, finite-dimensional subspace of $H_0^1(\Omega)^2$ and $\mathcal{M}$ be a finite-dimensional subspace of $H_0^1(\Omega)$. (Here $H_0^1(\Omega)^2$ is just the collection of 2-component functions which have each component in $H_0^1(\Omega)$.)

The continuous-time Galerkin approximation of this problem will be taken to be mappings of time into $\mathcal{V}$ and $\mathcal{M}$. The velocity **u** will be approximated by U, where $\mathsf{U}(t) \in \mathcal{V}$. When it is convenient we will also consider U to be a function of (x, y, t). The approximate temperature will have the form $\mathsf{T}(x, y, t) = \mathsf{S}(x, y, t) - (\Delta T/H)y$, where $\mathsf{S}(t) \in \mathcal{M}$. Let $(\cdot, \cdot)$ denote the inner product on $L^2(\Omega)^k$, for $k = 1, 2, 4$. We require that U and T satisfy the following conditions for all $t > 0$, $\varphi = (\varphi_1, \varphi_2) \in \mathcal{V}$, and $\psi \in \mathcal{M}$:

$$\nu(\nabla \mathsf{U}, \nabla \varphi) - (g\beta \mathsf{T}, \varphi_2) = 0, \quad (2.4)$$

$$(\partial_t \mathsf{T}, \psi) + (\mathsf{U} \cdot \nabla \mathsf{T}, \psi) + (D\nabla \mathsf{T}, \nabla \psi) = 0. \quad (2.5)$$

In the above $\nabla \mathsf{U}$ is viewed as the 4-vector consisting of the partial derivatives of the two components of U. To be sure that the notation is clear we explicitly note that

$$(\nabla \mathsf{U}, \nabla \varphi) = \int_\Omega [(\nabla \mathsf{U}_1 \cdot \nabla \varphi_1) + (\nabla \mathsf{U}_2 \cdot \nabla \varphi_2)] dx dy,$$

$$(D\nabla \mathsf{T}, \nabla \psi) = \int_\Omega [D\nabla \mathsf{T} \cdot \nabla \psi] dx dy.$$

Because the function $\varphi \in \mathcal{V}$ is divergence-free and vanishes on the boundary the $\nabla(p/\rho + gy)$ term drops out of the Galerkin relations. For definiteness we take $\mathsf{T}(0)$ to be the best $L^2(\Omega)$ fit to $T(0)$; this gives

$$\left(\mathsf{S}(0) - T(x, 0) - y\frac{\Delta T}{H}, \psi \right) = 0, \text{ for all } \psi \in \mathcal{M}. \quad (2.6)$$

The space of approximate solutions is called the space of trial functions because we try to find a solution in that space. In the case of a Galerkin method, the trial functions are also the so-called test functions; the differential equation is required to hold only in the sense that the residual, the amount by which the equations fail to hold, is chosen to be orthogonal to the test functions.

The equation (2.4) defines U as a linear function of T; putting it a little differently, it defines U as an affine function of S. We can then view (2.5) as a system of ordinary differential equations (ODEs) that define S, provided that S(0) is known. Since there is a quadratic nonlinearity in the resulting set of of ODEs, it is not clear, a priori, that the solution exists for all time. It is instructive to examine the stability of the system defined by (2.4) and (2.5).

Let $\|\cdot\|$ denote the $L^2(\Omega)$ norm. Take $\varphi = \mathsf{U}$ in (2.4) and $\psi = \mathsf{S}$ in (2.5) to see that

$$\nu\|\nabla \mathsf{U}\|^2 - (g\beta(\mathsf{S} - \frac{\Delta T}{H}y), \mathsf{U}_2) = 0,$$

$$\frac{1}{2}\frac{d}{dt}\|\mathsf{S}\|^2 + (\mathsf{U}\cdot\nabla(\mathsf{S} - \frac{\Delta T}{H}y), \mathsf{S}) + D(\nabla(\mathsf{S} - \frac{\Delta T}{H}y), \nabla \mathsf{S}) = 0.$$

Now note that $(\mathsf{U}\cdot\nabla\mathsf{S}, \mathsf{S}) = -\frac{1}{2}(\nabla\cdot\mathsf{U}, \mathsf{S}^2) = 0$, by integration by parts and the divergence free condition on $\mathcal{V}$. Also observe that $(\nabla y, \nabla \mathsf{S}) = (1, \partial_y \mathsf{S}) = 0$, since S vanishes on the boundary. Because U is divergence-free we get that $\int_0^L \mathsf{U}_2(x, y)dx = 0$ and this implies that $(y, \mathsf{U}_2) = 0$. Hence the displayed relations simplify to

$$\nu\|\nabla \mathsf{U}\|^2 - (g\beta \mathsf{S}, \mathsf{U}_2) = 0, \tag{2.7}$$

$$\frac{1}{2}\frac{d}{dt}\|\mathsf{S}\|^2 - (\mathsf{U}_2\frac{\Delta T}{H}, \mathsf{S}) + D\|\nabla \mathsf{S}\|^2 = 0. \tag{2.8}$$

If we add these two relations and use the fact that

$$\|\mathsf{U}_2\|^2 \leq \left(\frac{H}{\pi}\right)^2 \|\partial_y \mathsf{U}_2\|^2 \leq \frac{1}{2}\left(\frac{H}{\pi}\right)^2 \|\nabla \mathsf{U}\|^2, \tag{2.9}$$

then we can get

$$\frac{d}{dt}\|\mathsf{S}\|^2 + 2\nu\|\nabla \mathsf{U}\|^2 + 2D\|\nabla \mathsf{S}\|^2 \leq \sqrt{2}(g\beta + \frac{\Delta T}{H})\frac{H}{\pi}\|\nabla \mathsf{U}\|\|\mathsf{S}\|$$

$$\leq \nu\|\nabla \mathsf{U}\|^2 + C\|\mathsf{S}\|^2.$$

This implies that the solution S is bounded on any finite time interval. It then follows from (2.7) and (2.9) that U is bounded on any time finite time interval. Hence the solution exists for all positive time. Further, the norm of the solution can be bounded in terms of the time and the data of the problem, independently of the choice of subspaces $\mathcal{V}$ and $\mathcal{M}$.

A more careful version of the stability argument above gives that $\|\mathsf{S}\|$ decreases to zero if the Rayleigh number, Ra, satisfies

$$Ra = \frac{H^3 g\beta\Delta T}{D\nu} < 2\pi^4. \tag{2.10}$$

The manipulations are as follows. Add $\Delta T/(g\beta H)$ times (2.7) to (2.8), use the Cauchy inequality on the (S, U_2) term, and apply (2.9) and its analogue for S. This is a rather crude estimate of the critical Rayleigh number (it is off by a factor of approximately nine); however, we used almost no information about the nature of the subspaces.

To derive the reduced-dimension equations that we are seeking, we will use function spaces that have bases that are products of functions of x and y. We will take the limit of the permitted spaces in the x variable; this gives partial differential equations in x and t.

2.1 Reduction with One Temperature Trial Function

As a first example consider trial functions of the form

$$\mu(y) = \frac{1}{4}y^2(y-H)^2, \tag{2.11}$$

$$\mathsf{U} = \begin{pmatrix} \mathsf{U}(x,t)\mu_y(y) \\ -\mathsf{U}_x(x,t)\mu(y) \end{pmatrix}, \tag{2.12}$$

$$\varrho(y) = y(H-y), \tag{2.13}$$

$$\mathsf{T} = \mathsf{S}(x,t)\varrho(y) - \frac{\Delta T}{H}y, \tag{2.14}$$

where we have reused (overloaded if you will) the function names U and S and used subscripts x and y to denote differentiation. Note that for any differentiable function $\mathsf{U}(x,t)$ the the velocity field is divergence-free by construction.

With these trial functions the Galerkin method gives us a pair of equations

$$\partial_x^4 \mathsf{U} - \frac{24}{H^2}\partial_x^2 \mathsf{U} + \frac{504}{H^4}\mathsf{U} + \frac{18}{H^2}\frac{g\beta}{\nu}\partial_x \mathsf{S} = 0, \tag{2.15}$$

$$\partial_t \mathsf{S} - \frac{3H^2}{56}\frac{\Delta T}{H}\partial_x \mathsf{U} - D(\partial_x^2 \mathsf{S} - \frac{10}{H^2}\mathsf{S}) = 0. \tag{2.16}$$

To derive these equations one needs to evaluate several integrals which are recorded here for reference:

$$(\mu, \mu) = H^9 \frac{1}{10\,080},$$

$$(\mu_y, \mu_y) = -(\mu_{yy}, \mu) = H^7 \frac{12}{10\,080},$$

$$(\mu_{yy}, \mu_{yy}) = -(\mu_{yyy}, \mu_y) = H^5 \frac{504}{10\,080},$$

$$(\mu, \varrho) = H^7 \frac{-18}{10\,080},$$

$$(\varrho, \varrho) = H^5 \frac{336}{10\,080},$$

$$(\varrho_y, \varrho_y) = H^3 \frac{3,360}{10\,080},$$

$$(\mu_y\varrho, \varrho) = (\mu\varrho_y, \varrho) = 0.$$

In addition to satisfying the equations (2.15) and (2.16) the functions $U, \partial_x U$, and S vanish at $x = 0$ and $x = L$.

Notice that this is a linear system even though the system we started from was nonlinear. Further, it is easy to reduce this to a single equation for U; the equation has the form
$$L_4 \partial_t U + L_6 U = 0,$$
where the linear operators L_4 and L_6 are given by
$$\begin{aligned} L_4 w &= \partial_x^4 w - c_2 \partial_x^2 w + c_0 w, \\ L_6 w &= -b_6 \partial_x^6 w + b_4 \partial_x^4 w - b_2 \partial_x^2 w + b_0 w, \end{aligned}$$
where the b_i's and the c_i's are constants which depend on the coefficients in (2.15)–(2.16).

Although we will see that the Galerkin method tries to approximate the solution in a certain sense, the system produced above does not behave very much like the physical system or its Boussinesq approximation (2.1)–(2.3). The solutions of this system generically either decay to zero or grow unboundedly; for special values of the parameters a bounded steady solution can be obtained. This is in contrast to the boundedness of the solutions of the system we started from to get this Galerkin method, and serves as a cautionary note that this machine for producing reduced problems is not to be used without thought.

Physically the reason that the solution stays bounded is that as the fluid moves faster, less of it is heated at the bottom and cooled at the top; this means that there is less buoyant force to lift the hot fluid. In the approximate equations we have not given enough flexibility to allow the development of the local variation in temperature, so that an unbounded amount of energy can be put into the system by the imposed temperature difference.

2.2 Reduction with Two Temperature Trial Functions

Now consider a slightly richer set of trial functions for the temperature. With U as above, take S to have the following form
$$S(x, y, t) = S_1(x, t)\varrho(y) + S_2(x, t)\mu_y(y) - \frac{\Delta T}{H} y. \tag{2.17}$$

With these trial functions the Galerkin method gives us three partial differential equations:
$$\partial_x^4 U - \frac{24}{H^2} \partial_x^2 U + \frac{504}{H^4} U + \frac{18}{H^2} \frac{g\beta}{\nu} \partial_x S_1 = 0, \tag{2.18}$$

$$\partial_t S_1 - \frac{H^4}{168} U \partial_x S_2 - \frac{3H^2}{56} \frac{\Delta T}{H} \partial_x U$$
$$- \frac{H^4}{112} S_2 \partial_x U - D(\partial_x^2 S_1 - \frac{10}{H^2} S_1) = 0, \tag{2.19}$$

$$\partial_t S_2 - \frac{H^2}{6} U \partial_x S_1 + \frac{H^2}{12} S_1 \partial_x U - D(\partial_x^2 S_2 - \frac{42}{H^2} S_2) = 0. \tag{2.20}$$

In addition to this system the functions $U, \partial_x U, S_1,$ and S_2 vanish at $x = 0$ and $x = L$. In deriving the above equations we used the 1-dimensional inner products given before and a few more:

$$(\mu_y^2, \varrho) = H^9 \frac{-2}{10\,080},$$
$$(\mu\mu_{yy}, \varrho) = H^9 \frac{3}{10\,080},$$
$$(\mu\varrho_y, \mu_y) = H^9 \frac{-1}{10\,080},$$
$$(\mu\mu_{yy}, \mu_y) = (\mu_y^2, \mu_y) = 0.$$

2.3 Numerical Solution and Results

Although the equations (2.18)–(2.20) were derived from a Galerkin approach, these partial differential equations can be approximated using other techniques. We solved them numerically using a centered finite difference scheme. All variables were knot centered and fluxes were evaluated at the center of each interval using averaged quantities, e.g.,

$$U_{i+\frac{1}{2}} = \frac{1}{2}(U_{i+1} + U_i). \tag{2.21}$$

The fourth-order system is rewritten as a second-order system using a new variable, W, defined by

$$W = \partial_y^2 U. \tag{2.22}$$

In this case this choice was just for convenience; it allowed us to reuse software. However, in some other cases, involving singularity development and highly refined grids, we have found that the reduction from fourth-order to second improves the rounding behavior of the calculations.

Time stepping was based on a first-order backward difference scheme. The resulting nonlinear systems were solved using Newton's method with 1 or 2 iterations. The time step, Δt, was controlled by a step doubling technique. To advance time by Δt we take three backward Euler steps, one of size Δt and two of size $\Delta t/2$. The difference between the two approximations at end of the step is controlled as a way of assuring that the time truncation is small. Finally the two solutions are combined to eliminate first-order error terms in Δt. So if $C(\Delta t, k)$ is the solution at the end of k time steps of size $\Delta t/k$ of the backward difference scheme,

$$C_{\text{extrapolated}} = 2C(\Delta t, 2) - C(\Delta t, 1). \tag{2.23}$$

The resulting method is second-order correct in time.

Figure 1 shows some numerical results for the two temperature trial function Galerkin scheme. The parameter values for these results are as follows:

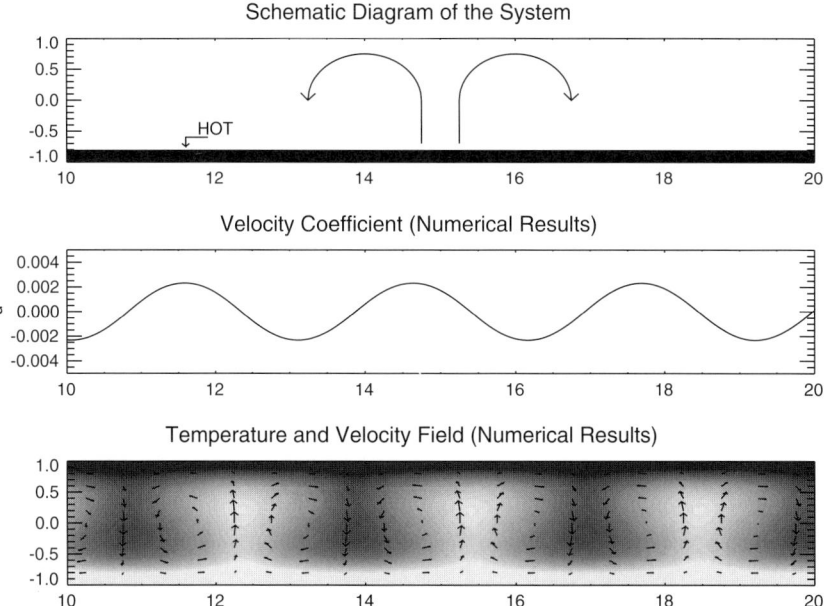

FIG. 1. These figures show numerical results for the two temperature trial function Galerkin scheme. The numerics were run for a domain 40 cm long and 2 cm high; we have shown a segment from the middle of the plates which is representative of the entire solution. The bottom plot shows the temperature and velocity fields between the two plates. The arrows show the convection rolls and color represents temperature; high temperature is light and low temperature is dark.

$$\begin{aligned} \Delta T &= 0.006 \text{K}, \\ \rho &= 1.0 \text{g/cm}^3, \\ \nu &= 0.01 \text{ cm}^2/\text{s}, \\ \beta &= 2.2 \times 10^{-4}/\text{K}, \\ D &= 5 \times 10^{-5} \text{ cm}^2/\text{s}. \end{aligned}$$

For future reference we note that the Rayleigh number (see (2.10)) in this case is about 20 700. The boundary conditions on the ends of the plate are $u = 0$ at $x = 0$ and $x = L$, $u' = 0$ at $x = 0$ and $x = L$, and no temperature flux across the walls. The shading in Fig. 1 was done by recombining the solution to equations (2.18)–(2.20) with the trial functions to produce a two-dimensional image.

2.4 Linear Stability of the Reduced System

One of the questions that has been well studied for this system is the onset of convection. If the fluid is at rest and the temperature difference is small, the fluid will remain at rest, but if the temperature difference is increased, rolls develop that have a characteristic wave-length. We can easily study the stability of the no-flow solution for the reduced system to see how well it corresponds to the full system.

The linearized equations for the system using two trial functions for temperature is (using the same names for the variables as perturbations in those variables)

$$\partial_x^4 U - \frac{24}{H^2}\partial_x^2 U + \frac{504}{H^4}U + \frac{18}{H^2}\frac{g\beta}{\nu}\partial_x S_1 = 0, \qquad (2.24)$$

$$\partial_t S_1 - \frac{3H^2}{56}\frac{\Delta T}{H}\partial_x U - D(\partial_x^2 S_1 - \frac{10}{H^2}S_1) = 0, \qquad (2.25)$$

$$\partial_t S_2 - D(\partial_x^2 S_2 - \frac{42}{H^2}S_2) = 0. \qquad (2.26)$$

The convenient fact is that the variable S_2 is completely separate from the other two and is always stable. Thus we need only consider the pair U and S_1. First, write a single equation for U by differentiating the S_1 equation (2.25) with respect to x and using the U equation (2.24) to replace $\partial_x S_1$. This process gives

$$\partial_t \mathcal{L}_4 U + \mathcal{L}_6 U = 0,$$

where

$$\mathcal{L}_4 w = \left(\partial_x^4 - \frac{24}{H^2}\partial_x^2 + \frac{504}{H^4}\right)\frac{H^2 \nu}{18g\beta}w,$$

$$\mathcal{L}_6 w = \frac{3H^2}{56}\frac{\Delta T}{H}\partial_x^2 - D(\partial_x^2 - \frac{10}{H^2})\mathcal{L}_4 w.$$

In looking at the stability of this system we will ignore the boundary conditions at $x = 0$ and $x = L$; in effect we are considering the case in which L is very large with respect to the wave-length of the unstable perturbation. We look for solutions of the form

$$U(x,t) = e^{\gamma t} e^{ikx/H}.$$

Notice that the operators $\mathcal{L}_4$ and $\mathcal{L}_6$ act on U by multiplication by a polynomial in k, and the action of $\mathcal{L}_4$ gives a positive multiple for any real k. From this we conclude that if the action of $\mathcal{L}_6$ is a negative multiplier, then the γ will be positive, and the solution will grow exponentially. Solutions with positive γ we call unstable.

The action of $\mathcal{L}_6$ is as follows:

$$\mathcal{L}_6 U = \left\{-\frac{3H^2}{56}\frac{\Delta T}{H}\left(\frac{k}{H}\right)^2\right.$$

$$+ D\left(\left(\frac{k}{H}\right)^2 + \frac{10}{H^2}\right)\left(\left(\frac{k}{H}\right)^4 + \frac{24k^2}{H^4} + \frac{504}{H^4}\right)\frac{H^2\nu}{18g\beta}\Bigg\}\mathsf{U}$$

$$= \frac{D\nu}{18H^4 g\beta}\left\{(k^2+10)(k^4+24k^2+504) - \frac{27}{28}\frac{\Delta T g \beta H^3}{D\nu}k^2\right\}\mathsf{U}.$$

Thus $\mathcal{L}_6 \mathsf{U}$ is a negative multiple of U if

$$p_3(k^2) = (k^2+10)(k^4+24k^2+504) - \frac{27}{28}Rak^2 < 0,$$

where the Rayleigh number, Ra, is given by (2.10). Thus we are interested in whether or not p_3 is negative at any point on $(0, \infty)$. Note that if $p_3'(0) \geq 0$ then p_3 is positive and monotone increasing on $(0, \infty)$. However, if $p_3'(0) < 0$ then it has exactly one minimum on $(0, \infty)$ and the location of that minimum is

$$\frac{34}{3} + \sqrt{\left(\frac{34}{3}\right)^2 + \frac{9}{28}Ra - 248}.$$

This minimum is positive for small Ra and negative for large Ra. The transition is the critical value $Ra_{crit} \simeq 1749.98$. The corresponding value of k^2 is $9.71276 \simeq (3.1165)^2$. The critical wave-number of 3.1165 gives a wave-length of $2\pi H/3.1165$. If one does the stability analysis on the original system of multidimensional partial differential equations, the critical Rayleigh number is computed to be 1708 and the wave-number is $3.13/H$ [3]. A simulation very close to that for Fig. 1 was run with ΔT reduced so that $Ra = 1776$ was just above the critical value. This gave a wave length of 3.9cm, which translates to a wave-number of $3.2/H$. At $Ra = 1776$ the above analysis gives that it should be $3.4/H$; this discrepancy seems reasonable to us since the simulation has zero boundary conditions at $x = 0$ and $x = 40$.

2.5 Error Bound for Galerkin Approximations

In showing that a Galerkin method converges, one frequently shows that the error in the Galerkin solution is bounded by a multiple of an approximation error term. Loosely speaking, the result says that the Galerkin solution is a good approximation if there is a good approximation possible in the function space being used. This is the approach that we will use here.

Take $\mathsf{W}(t)$ and $\mathsf{Z}(t)$ to be smooth functions into $\mathcal{V}$ and $\mathcal{M}$, respectively. Whenever it is convenient we will consider W and Z to be functions of (x, y, t). Let $s(x, y, t) = T(x, y, t) + y\Delta T/H$ and adopt the notation

$$\vartheta = \mathsf{W} - \mathsf{U},$$
$$\theta = \mathsf{Z} - \mathsf{S},$$
$$\eta = \mathsf{W} - \mathbf{u},$$
$$\zeta = \mathsf{Z} - s.$$

The plan of attack is to show that ϑ and θ are small if η and ζ are; the triangle inequality will then imply that the errors $\mathbf{u} - \mathbf{U}$ and $s - \mathsf{S} = T - \mathsf{T}$ are small if η and ζ are.

Note that for all $\varphi \in \mathcal{V}$ and $\psi \in \mathcal{M}$ we have

$$\nu(\nabla \vartheta, \nabla \varphi) = \nu(\nabla \eta, \nabla \varphi) + g\beta(\theta - \zeta, \varphi_2),$$
$$(\partial_t \theta + \mathbf{U} \cdot \nabla \theta, \psi) + D(\nabla \theta, \nabla \psi) = (\partial_t \zeta + (\eta - \vartheta) \cdot \nabla T + \mathbf{U} \cdot \nabla \zeta, \psi)$$
$$+ D(\nabla \zeta, \nabla \psi).$$

Make the choice of W and Z such that they match $\mathbf{u}$ and s as closely as possible in the norm given by taking the L^2-norm of the gradient. Then for all $\varphi \in \mathcal{V}$ and $\psi \in \mathcal{M}$,

$$(\nabla \eta, \nabla \varphi) = 0,$$
$$(\nabla \zeta, \nabla \psi) = 0.$$

This then gives

$$\nu(\nabla \vartheta, \nabla \varphi) = g\beta(\theta - \zeta, \varphi_2), \tag{2.27}$$
$$(\partial_t \theta + \mathbf{U} \cdot \nabla \theta, \psi) + D(\nabla \theta, \nabla \psi) = (\partial_t \zeta + (\eta - \vartheta) \cdot \nabla T, \psi)$$
$$- (\mathbf{U} \zeta, \nabla \psi). \tag{2.28}$$

In (2.27) use $\varphi = \vartheta$ and in (2.28) use $\psi = \theta$ to see that

$$\|\nabla \vartheta\|^2 = g\beta(\theta - \zeta, \vartheta_2) \tag{2.29}$$
$$\frac{1}{2} \frac{d}{dt} \|\theta\|^2 + D \|\nabla \theta\|^2 = (\partial_t \zeta + (\eta - \vartheta) \cdot \nabla T, \theta)$$
$$- (\mathbf{U} \zeta, \nabla \theta). \tag{2.30}$$

In deriving (2.30) we used the fact that $(\mathbf{U} \cdot \nabla \theta, \theta) = 0$, which follows since $\theta \nabla \theta = \frac{1}{2} \nabla(\theta^2)$ and $\mathbf{U}$ is divergence-free.

In order to be able to use these two relations to bound θ and ϑ we need to bound the terms on the right-hand sides. Many of the terms will involve the use of the fact that for any $c > 0$ and any real numbers a and b,

$$ab = \varepsilon a^2 + \frac{1}{4\varepsilon} b^2 - (\sqrt{\varepsilon} a - \frac{1}{2\sqrt{\varepsilon}} b)^2$$
$$\leq \varepsilon a^2 + \frac{1}{4\varepsilon} b^2.$$

To simplify the calculations we will use C to denote a generic constant, not necessarily the same at each occurrence, which does not depend on the approximate solution. For example instead of writing

$$g\beta(\theta - \zeta, \vartheta_2) \leq g\beta \|\theta - \zeta\| \frac{H}{\pi} \|\partial_y \vartheta_2\|$$

$$\leq \frac{g\beta H}{\sqrt{2\pi}}\|\theta-\varsigma\|\|\nabla\vartheta\|$$

$$\leq \frac{1}{4}\|\nabla\vartheta\|^2 + \frac{1}{2}\left(\frac{g\beta H}{\pi}\right)^2\|\theta-\varsigma\|^2$$

$$\leq \frac{1}{4}\|\nabla\vartheta\|^2 + \left(\frac{g\beta H}{\pi}\right)^2(\|\theta\|^2+\|\varsigma\|^2),$$

we will simply say

$$g\beta(\theta-\varsigma,\vartheta_2) \leq \frac{1}{4}\|\nabla\vartheta\|^2 + C\left(\|\theta\|^2+\|\varsigma\|^2\right).$$

We will need the fact that

$$(\partial_t\varsigma,\theta) \leq \|\partial_t\varsigma\|^2 + \|\theta\|^2.$$

We assume for this discussion that ∇T and $\mathbf{u}$ are uniformly bounded, so that in particular

$$((\eta-\vartheta)\cdot\nabla T,\theta) \leq \frac{1}{8}\|\nabla\vartheta\|^2 + C\left(\|\theta\|^2+\|\eta\|^2\right).$$

The term $(\mathsf{U}\varsigma,\nabla\theta)$ is a little more complicated. First we use

$$-(\mathsf{U}\varsigma,\nabla\theta) \leq \frac{D}{2}\|\nabla\theta\|^2 + C\|\mathsf{U}\varsigma\|^2.$$

We know from the bound that we got for the Galerkin solution that on any finite time interval U is bounded in $H^1(\Omega)$. Thus,

$$\|\mathsf{U}\varsigma\|^2 \leq \|\mathsf{U}\|_{L^4}^2\|\varsigma\|_{L^4}^2 \leq C\|\varsigma\|_{H^1}^2;$$

one can use weaker norms with more complicated arguments. We are using the fact that the norm on $L^4(\Omega)$ can be bounded by a multiple of the norm on $H^1(\Omega)$.

Let $\|\phi\|_1$ denote the $H^1(\Omega)$ norm of ϕ. With the bounds derived above we can add (2.29) and (2.30) to get

$$\frac{d}{dt}\|\theta\|^2 + D\|\nabla\theta\|^2 + \|\nabla\vartheta\|^2 \leq C\left[\|\theta\|^2 + \|\partial_t\varsigma\|^2 + \|\varsigma\|_1^2 + \|\eta\|^2\right].$$

By Gronwall's inequality this implies that for any bounded interval $J = (0, t_{final})$ there exists a constant C such that

$$\max_{t\in J}\|\theta(t)\|^2 + \int_J(D\|\nabla\theta\|^2+\|\nabla\vartheta\|^2)dt$$

$$\leq C\left[\|\theta(0)\|^2 + \int_J(\|\eta\|^2+\|\varsigma\|_1^2+\|\partial_t\varsigma\|^2)dt\right]$$

$$\leq C\left[\|\varsigma(0)\|^2 + \int_J(\|\eta\|^2+\|\varsigma\|_1^2+\|\partial_t\varsigma\|^2)dt\right]. \quad (2.31)$$

In the last step above we used the fact that $\|\theta(0)\| \leq \|\zeta(0)\|$; this follows because we took $S(0)$ to be the $L^2(\Omega)$ projection of $s(0)$.

For a function defined on $\Omega \times J$ let $L^p(H^k)$ denote the $L^p(J)$ norm of its $H^k(\Omega)$ norm. It is not hard to show that

$$\|\zeta\|^2_{L^\infty(L^2)} \leq C \|\zeta\|^2_{L^2(L^2)} + \|\partial_t \zeta\|^2_{L^2(L^2)}.$$

(One can take the C to be $1 + 1/t_{final}$.) The relation (2.31) together with the triangle inequality give the following result.

Theorem 2.1 *Suppose that the solution* $(\mathbf{u}, T)$ *is such that* $\mathbf{u}$, T *and* ∇T *are bounded uniformly on* $\Omega \times J$, *where* $J = (0, t_{final})$. *Then there is a constant C such that if* U *and* T *are the Galerkin approximations defined by (2.4)–(2.6),*

$$\|T - \mathsf{T}\|_{L^\infty(L^2)} + \|T - \mathsf{T}\|_{L^2(H^1)} + \|\mathbf{u} - \mathsf{U}\|_{L^2(H^1)}$$
$$\leq C \left[\|\zeta\|_{L^2(H^1)} + \|\partial_t \zeta\|_{L^2(L^2)} + \|\eta\|_{L^2(H^1)} \right].$$

3 Colloidal Shocks

In this section we give a summary of the result of applying techniques like those described above to a situation in which colloids are seen to spontaneously form shocks. This work, which is discussed in detail in [7], was inspired by the experiments reported in [9], but observations of this effect go back at least to 1884 [2].

In the experiments of [9], tubes about 30 cm tall and 2 cm in diameter were filled with a suspension of beads in water that were about 10^{-4}cm in diameter and had a density of about 0.73 g/cm^3. After being left for days the tubes had sharply delineated bands of differing concentration from top to bottom. Careful observation revealed that these bands would disappear, but would then reform in the same locations later. A line of dye down the center of the cylinder deformed in such a way as to indicate that there was a convection roll within each band.

3.1 Differential Model

Instead of using a tube we will initially discuss a model for a slab of fluid between two parallel vertical plates located at $x = \pm d$ (see Fig. 2). Suppose that the plates are each held at uniform temperature and that there is a temperature difference ΔT between the plates. We will assume that the fluid flow has essentially no effect on the temperature, so that we can take the temperature field to be $T(x, y) = T_0 + x\Delta T/2d$. In the limit of ΔT going to zero, the speed of the fluid goes to zero, but the thermal relaxation time for the domain stays fixed; thus the assumed behavior for temperature is reasonable for small ΔT. Let $c = c(x, y, t)$ denote the volume fraction of the fluid that is beads and take $\Delta \rho$ to be the density of the water less the density of the beads. Then we can write a momentum balance equation that is similar in spirit to the one used in the previous section:

$$\frac{\nabla p}{\rho} - \nu \nabla \cdot \nabla \mathbf{u} + \frac{\Delta \rho}{\rho} c \mathbf{g} + x\beta \frac{\Delta T}{2d} \mathbf{g} = 0. \tag{3.1}$$

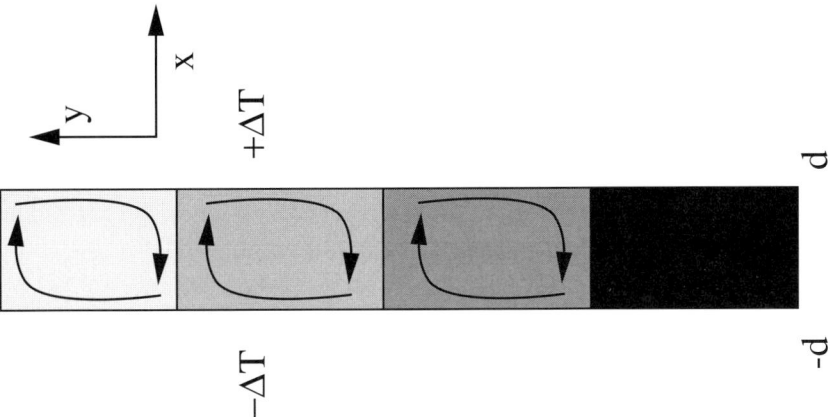

FIG. 2. This is a schematic diagram of the layers formed in a colloidal system with a lateral temperature gradient. The equations we use are two-dimensional so this represents a cross-section between two semi-infinite planes.

The boundary conditions for **u** are that both components vanish on the boundary. If $c \equiv 0$ and we neglect end effects then the solution to this problem is

$$u_1 \equiv 0, \quad u_2 = -\frac{g\beta\Delta T}{12\nu d}\left(x^3 - xd^2\right).$$

The conservation of beads equation has the form

$$\partial_t c + \mathbf{u} \cdot \nabla c + \mathbf{v}_0 \cdot \nabla(c - kc^2) - D\nabla \cdot \nabla c = 0, \tag{3.2}$$

where $\mathbf{v}_0 = 2a^2\Delta\rho g/9\rho\nu$ is the velocity of an isolated particle of radius a when buoyancy is balanced by Stokes drag [8], D is the diffusivity of the beads, and k is a constant. Batchelor showed [1] that for colloids small enough to be kept randomly distributed by thermal fluctuations, $k \sim 6.55$; k quantifies the influence that the motion of one particle, which induces a flow, has on the other particles. Solutions, in which the particles don't move due to buoyancy, have $\mathbf{v}_0 = 0$. The boundary conditions for c are that there is no flux through the walls of the domain.

3.2 Galerkin Reduction

In the treatment of Rayleigh–Bénard convection we arrived at a partial differential equation (PDE) in (x, t). Here we will use Galerkin to treat the variations in x and arrive at a PDE in (y, t). The reduced solutions are of the form

$$\mathsf{U}(x, y, t) = \begin{pmatrix} -\mathsf{U}_y(y, t)\mu(x) \\ \mathsf{U}(y, t)\mu_x(x) \end{pmatrix},$$

$$\mu(x) = \frac{1}{4}(x^2 - d^2)^2,$$
$$C(x, y, t) = \bar{C}(y, t) + B(y, t)\lambda(x),$$
$$\lambda(x) = x(x^2 - 3d^2).$$

Note that the trial functions for C satisfy $\partial_x C = 0$ at $x = \pm d$. Just as in the previous section, the approximate velocity is divergence-free by construction. Using the natural Galerkin method for these trial functions we get the following set of reduced equations:

$$\frac{1}{3}\partial_y^4 U - 2\partial_y^2 U + \frac{21}{2d^2}U - \frac{g}{2\nu}\left(\frac{9\Delta\rho}{\rho}B - \frac{7\beta\Delta T}{4d^3}\right) = 0,$$

$$\partial_t \bar{C} + \partial_y F - D\partial_y^2 \bar{C} = 0,$$

$$\partial_t B + \frac{42D}{17d^2}B + \partial_y G + \frac{3}{17}U\partial_y \bar{C} - D\partial_y^2 B = 0,$$

where the fluxes F and G are given by

$$F = v_0(\bar{C} - k\bar{C}^2) + \frac{4}{35}d^6(3UB - \frac{17}{2}kv_0 B^2),$$
$$G = v_0 B(1 - k\bar{C}).$$

3.3 Computational Results

The experimental work of [11] showed that the shocks form much more quickly if there is a gradient in concentration, so we chose to simulate a case in which the initial conditions had a uniform gradient in concentration of beads; the concentration went from approximately 0.85×10^{-4} at the bottom to 1.1×10^{-4} at the top. This concentration gradient should be stable if $\Delta T = 0$ since the heavier fluid is below the lighter fluid. The parameters were taken to be $d = 0.4$ cm, $H = 25$ cm, $\nu = 0.01$ cm^2/s, and $\beta = 2.2 \times 10^{-4}$ K^{-1}, (the values for ν and β correspond to water), $D = 5 \times 10^{-8}$ cm^2/s, $v_0 = 1.323 \times 10^{-6}$ cm/s, and $\Delta T = 0.005$K.

The time sequences for concentration and speed show that the rolls form first at the top and bottom; these rolls propagate into the "tube" and eventually shed new rolls. The behavior of the solution can be described in a very intuitive way once we have seen the solutions.

First consider what happens far from the ends of the "tube". When the temperature differential is first applied, a large convection roll starts to form; however, the fluid moves only far enough for the density change due to concentration of beads to cancel the density change due to the temperature. If we neglect end effects, this works fine since the initial concentration gradient gave us heavier fluid at the bottom. One can easily quantify the distance the fluid has to roll to balance the density at each level. Notice that although we come back to a state where the density is constant in each horizontal plane, there is a concentration gradient in those planes.

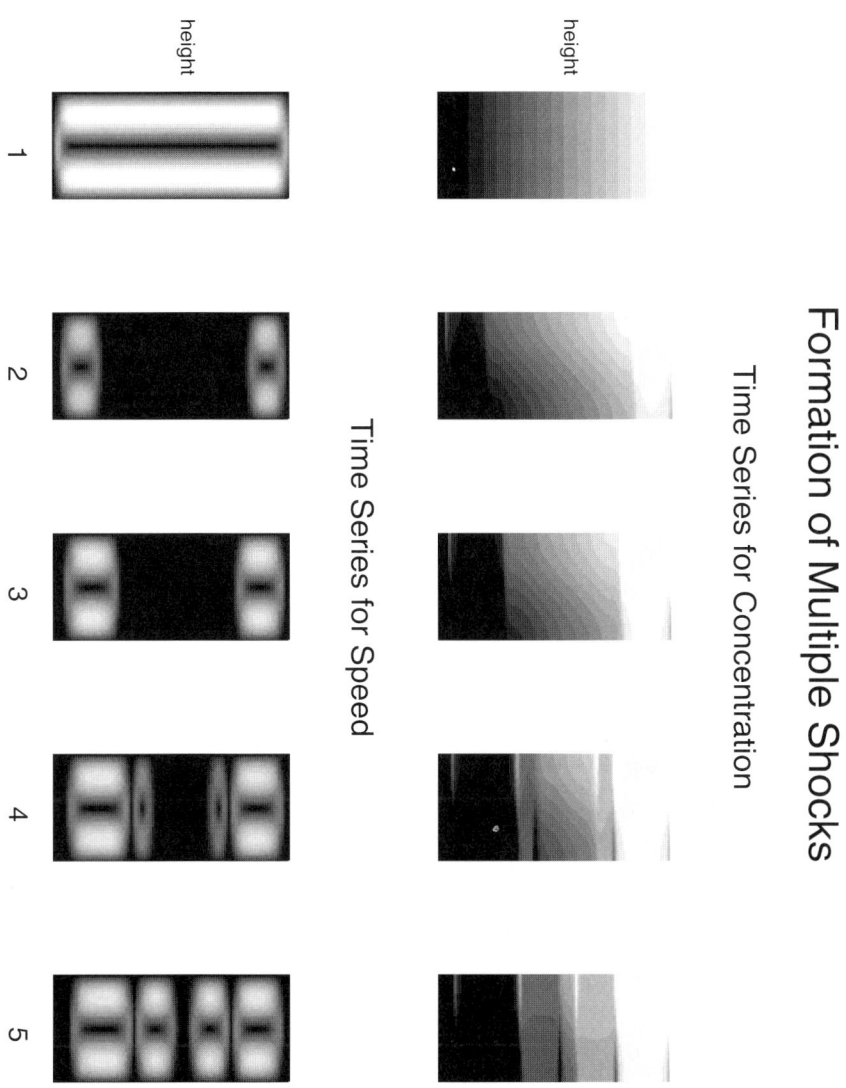

FIG. 3. These plots show a time series evolution for concentration and speed. In the first series, shading represents concentration; high concentrations are shown with light shading and low with dark. In the second series, shading represents the magnitude of the velocity; light is fast and dark is slow. We can clearly see layers form as rolls are shed off the top and bottom of the "tubes."

However, the fluid at the top and bottom does something very different than in the intermediate regions. At the ends the flow is mostly horizontal and hence does not bring in fluid of different concentration to balance the thermal effect. Thus the fluid continues to roll in small cells at the top and bottom. This has the result of homogenizing these regions so that the rolls continue. But viscosity causes the rolls to expand by dragging some of the almost quiescent fluid in the middle. The speed of the advancing boundary of the cell slows as the cell grows, because the fluid being entrained is increasingly different in concentration from the stirred fluid in the cell.

Finally, there is the effect of horizontal diffusion of beads. Because of this, the fluid in the quiescent region must continue to rotate very slowly. After the advance of the top and bottom cells has slowed enough, this very slow roll in the big still region kicks off new cells at the interface between the still and moving regions. Once started, these cells roll at almost the speed of the steady $c \equiv 0$ solution discussed above, because the flow reduces the vertical concentration variation. Hence these relatively fast moving cells grow.

3.4 Alternative Equations

The reduced equations above were derived using a Galerkin approach, but there is a collection of other methods that could be used with the same trial functions. In [7] we give formulations that come from collocation and control volume methods. Collocation methods are based on requiring that the original differential equations hold at certain points in the domain, and control volume methods are based on requiring that certain moments of the residual vanish. It is not surprising to see that these various techniques give quite similar sets of reduced equations.

3.5 Round Tubes

In the foregoing discussion we used infinitely wide flat tubes to get an initial problem that had two space dimensions which we then reduced to a single space dimension. At first blush it might seem that one could use cylindrical symmetry to make a similar step in round tubes. However, if the external temperature field is linear in x while the tubes are cylindrical, the problem is more thoroughly three-dimensional. In this case one can again pose reasonable one space dimensional PDE's.

The first step is to find solution to the Stokes equation in an infinite cylinder with a forcing term that is linear in a fixed transverse direction. This vertical velocity is then used to build the horizontal terms, just as in the case of parallel plates. Think of the horizontal plane as being parametrized by (x, y) and take the vertical direction to be z. For convenience take the radius of the cylinder to be 1. With $r^2 = x^2 + y^2$ define the following functions:

$$\begin{aligned} \varphi^x(x,y) &= 5(1-r^2)^2 + 4y^2(1-r^2), \\ \varphi^y(x,y) &= -4xy(1-r^2), \end{aligned}$$

$$\varphi^z(x,y) = 24x(1-r^2).$$

Then for any differentiable function $f(z)$ the vector field

$$(f'\varphi^x, f'\varphi^y, f\varphi^z)$$

is divergence-free; these are the velocity trial functions in the Galerkin method for the reduced equations.

There are two concentration trial functions. Take

$$\lambda(x,y) = x(x^2 + y^2 - 3).$$

The concentration is then assumed to have the form

$$\mathsf{C}(x,y,z,t) = \mathsf{C}(z,t) + \mathsf{B}(z,t)\lambda(x,y).$$

The function λ has a zero normal derivative on the boundary of the tube. This is easily seem by using a polar representation $(x,y) = (r\cos(\theta), r\sin(\theta))$. In this form $\lambda(x,y) = \cos(\theta)r(r^2 - 3)$. The derivative with respect to r is the normal derivative and that function vanishes when $r = 1$.

These functions then give a one space dimension set of reduced equations for round tubes.

4 Drops and Jets

In [5] a reduction was presented for the Navier–Stokes equations in the context of axisymmetric flows that has something of the flavor of the previous two sections; that was in fact the work that caused us to look at these questions in this way. In [5] the technique involved an expansion in the aspect ratio rather than something that would be viewed as a numerical method. Here we will discuss the use of the Galerkin approach again but using a basis that is suited to axisymmetric flows in slender geometries. This work has been described in more detail in [4], where many connections to other work in hydrodynamics are given.

We first describe the partial differential equations and the appropriate weak form using Cartesian coordinates. We consider the case in which the fluid is in a domain $\Omega = \Omega(t)$ and restrict attention to the situation in which the entire boundary, $\Gamma = \partial\Omega$, is a free surface; the treatment of the case in which part of the boundary is fixed is frequently not difficult in practice, but it complicates the description. The equations that we want to solve are

$$\nabla \cdot \mathbf{u} = 0,$$
$$\partial_t \mathbf{u} + \mathbf{u} \cdot \nabla \mathbf{u} - \frac{1}{\rho}\nabla\sigma = 0,$$

where

$$\sigma_{ij} = -p\delta_{ij} + 2\eta D_{ij}(\mathbf{u}),$$

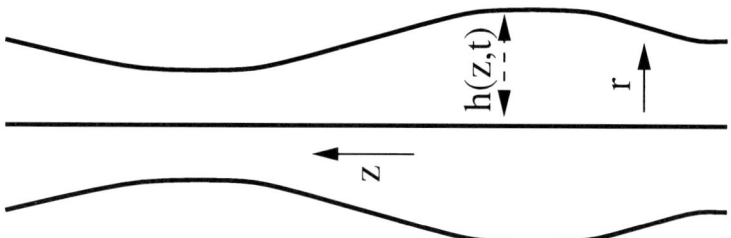

FIG. 4. This figure indicates the generic geometry that we have in the case of axisymmetric free surface flows.

and for a vector field, $\mathbf{w}$, $\mathbf{D}(\mathbf{w})$ is the strain rate tensor

$$D_{ij}(\mathbf{w}) = \frac{1}{2}(\partial_{x_j} u_i + \partial_{x_i} u_j).$$

Here η is the absolute viscosity as contrasted with the kinematic viscosity $\nu = \eta/\rho$ that we used in previous sections. We assume that η, ρ, and the surface tension γ are constant.

One must choose the weak form for the Navier–Stokes equations with some care to allow for the imposition of the free surface conditions [6]. The weak form of this set of PDE's that is useful here is as follows. Find $\mathbf{u}$ which is divergence-free such that for all smooth divergence-free functions $\mathbf{w}$,

$$\rho \int_\Omega [\partial_t \mathbf{u} + \mathbf{u} \cdot \nabla \mathbf{u}] \cdot w + \gamma \int_\Gamma \kappa \mathbf{n} \cdot \mathbf{w} + 2\eta \int_\Omega \mathbf{D}(\mathbf{u}) : \mathbf{D}(\mathbf{w}) = 0,$$

where κ is the mean curvature and $\mathbf{n}$ is the outward normal to Γ. In this equation $\mathbf{D}(\mathbf{u}) : \mathbf{D}(\mathbf{w})$ means the sum over all i, j of the products of the components.

We now work in (r, z) coordinates where $r = 0$ is the axis of symmetry, z measures distance along that axis, r measures the distance from the axis. The distance of the free surface from the axis will be denoted $h(z, t)$, and we will assume that the surface is such that h is single-valued; the geometry is illustrated in Fig. 4. We also assume that the velocity is only in the r and z directions. The functions that define a solution of the Navier–Stokes equations are then h and $\mathbf{u}$, where

$$\mathbf{u} = \begin{pmatrix} u^r \\ u^z \end{pmatrix}.$$

An easy way to build divergence-free functions in this context is to look for functions

$$\mathsf{U} = \begin{pmatrix} \mathsf{U}^r \\ \mathsf{U}^z \end{pmatrix},$$

where

$$\mathsf{U}^z(z, r, t) = \sum_{k=0}^{K} \mathsf{V}^k(z, t) r^{2k}.$$

It then follows that

$$U^r(z,r,t) = \sum_{k=0}^{K} -\partial_z V^k(z,t) r^{2k+1}/(2k+2),$$

since the desired condition is

$$\partial_r(rU^r) + r\partial_z U^z = 0.$$

To get the simplest possible approximation we restrict attention to the case $K = 0$; this gives (dropping the index k)

$$\mathsf{U} = \begin{pmatrix} -\frac{1}{2}rV(z,t) \\ \partial_z V(z,t) \end{pmatrix}.$$

Then the system that is given by the method described is

$$h^2 \partial_t V - \frac{1}{8}\partial_z(h^4 \partial_z \partial_t V)$$
$$+ h^2 V \partial_z V + \frac{1}{16}\partial_z(h^4 (\partial_z V)^2 - 2V\partial_z^2 V)$$
$$+ \frac{\gamma}{\rho} h^2 \partial_z \kappa - \nu \left(3\partial_z(h^2 \partial_z V) - \frac{1}{8}\partial_z^2(h^4 \partial_z^2 V) \right) = 0,$$
$$\partial_t(h^2) + \partial_z(h^2 V) = 0.$$

We can express the curvature as follows:

$$\kappa = \frac{1}{h\sqrt{1+(\partial_z h)^2}} - \frac{\partial_z^2 h}{(1+(\partial_z h)^2)^{3/2}}.$$

These equations are called the Cosserat equations and had previously been derived by requiring certain symmetries to be preserved by the approximate solution; see [4] for a discussion. Before the Galerkin derivation given in [4] it was not known that these equations could be derived directly from Navier–Stokes. The equations for the next-higher-order expansion are also given in [4].

In [5] the reduced equations are similar to those given above. The only difference is that the first equation is replaced by

$$h^2 \left(\partial_t v + v \partial_z v + \frac{1}{\rho}\partial_z p \right) - 3\nu \partial_z(h^2 \partial_z v) = 0,$$

where with κ as above,

$$p = \gamma \kappa.$$

Each of the terms in the Cosserat equations which are not in this simpler form are higher-order in h, so we expect that for very slender bodies they should give almost the same behavior. An interesting application of the equations of [5] to drop break-off can be found in [10].

5 Other Resources

Additional graphics and animations can be found at:

http://www.cs.uchicago.edu/~hosoi.

We expect that these pages will be maintained for at least five years.

Acknowledgements

This work was partially supported by the Office of Naval Research under ONR-AASERT N00014-94-1-0798, by the Department of Energy under DE-FG02-92ER25119, and by the National Science Foundation under DMR-9400379. Additionally, we would like to gratefully thank L. Kadanoff, M. Brenner, D. Grier, J. Eggers, R. Rosner, and L. Mahadevan for their valuable conversations.

Bibliography

1. Batchelor, G. K. (1972). *J. Fluid Mech.*, **52**, 245.
2. Brewer, W. H. (1884). *Mem. Nat. Acad. Sci. USA*, **2**, 165.
3. Chandrasekhar, S. (1957). Rumford Medal lecture 1957, thermal convection. *Daedalus*, **86**, 323–339. Reprinted in *Selected Papers S. Chandrasekhar* **4**. The University of Chicago Press, 163–191.
4. Eggers, J. (1997). Nonlinear dynamics and breakup of free-surface flows. *Reviews of Modern Physics*, **69**, pp 865-929.
5. Eggers, J. and Dupont, T. F. (1994). Drop formation in a one-dimensional approximation of the Navier–Stokes equation. *Journal of Fluid Mechanics*, **262**, 205–221.
6. Fortin, M. (1993). Finite element solution of the Navier–Stokes equations. *Acta Numerica 1993*, Cambridge Univ. Press, 239–284.
7. Hosoi, A. E. and Dupont, T. F. (1996), Layer formation in monodisperse suspensions and colloids, *Journal of Fluid Mechanics*, **328**, 297-312.
8. Landau, L. D. and Lifshitz, E. M. (1987). *Fluid Mechanics*, 2nd ed. Pergamon.
9. Mueth, D. M., Crocker, J. C., Esipov, S. E., and Grier, D. G. (1996). *Phys. Rev. Lett.*, **77**, 12.
10. Shi, X. D., Brenner, M. P., and Nagel, S. R. (1994). A cascade of structure in a drop falling from a faucet. *Science*, **265**, 157.
11. Siano, D. B. (1979). *J. of Colloid and Interface Sci.*, **68**, 111.
12. Tritton, D. J. (1988). *Physical Fluid Dynamics*, 2nd ed. Oxford.

5
Viscous Approximation to Transonic Gas Dynamics: Flow Past Profiles and Charged-Particle Systems

I. M. Gamba and C. S. Morawetz
Courant Institute of Mathematical Sciences, New York University

Abstract

A boundary value problem in a domain Ω is considered for a system of equations of fluid-Poisson type, i.e., a viscous approximation to a potential equation for the velocity, coupled with an ordinary differential equation along the streamlines for the density, and a Poisson equation for the electric field. A particular case of this system is a viscous approximation of transonic flow models. The general case is a model for semiconductors. We present an overview of the problem and, in addition, we show an improvement of the lower bound for the density that controls the rate of approach to cavitation density by a quantity of the order of the viscosity parameter raised to the power that corresponds to the inverse of the enthalpy function. This is a necessary step in the existing programs showing existence of a solution for the transonic flow problem.

1 Introduction

The present chapter deals with a steady-state fluid level model that is an approximation to inviscid potential flow that changes type. Viscosity and friction are added; existence of two dimensional solutions has been established by the authors (see [25]), along with some uniform bounds in the viscosity parameters, for geometries and boundary conditions corresponding to flow in channels that includes gas flow past a profile and charged particle transport in the modeling of semiconductor devices.

These models also appear in higher hierarchies of macroscopic approximation of charged-particle systems in the modeling of electron–ion plasmas and semiconductor devices, where the transport is induced by the superposition of an internal and an externally applied electric field.

The resulting macroscopic approximation yields a fluid level equation, coupled with a Poisson equation for the corresponding electric potential (see, for instance, Anile and Muscato [1], Azoff [3], Baccarani and Wordeman [4], Bløtekjaer [7], Bringer and Schön [8], Jerome [33], Markowich, Ringhofer, and Schmeiser [38], Poupaud [47], for justifications for these models). In addition, Section 3 contains a brief survey about mathematical developments on charged-particle transport in the modeling of semiconductor devices. Here, we "model a model" by tailoring the steady-state electro-hydrodynamic model into a solvable problem. By solvable, we mean we have established in [25] an existence theorem for a solution. The implication of the theorem is that such a model can be consistently computed.

The "adjusted" model, which has potential gas flow as a special case, hopefully gives insight into certain semiconductor regimes, just as potential transonic flow did for the full fluid equations with small viscous effects. Indeed, if no electric field is present, the system reduces to a two-dimensional, steady, irrotational, compressible, viscous flow model in a "channel". Classical references on this model can be found in Courant-Friedrichs [15], Morawetz [40], Serrin [49], and Synge [51].

1.1 Transonic Flow Equations

We recall that the transonic flow model is given by

$$\text{div}(\rho \nabla \varphi) = 0, \quad \frac{1}{2}|\nabla \varphi|^2 + i(\rho) = K, \tag{1.1}$$

where φ is the potential flow function, $\nabla \varphi$ the associated velocity field, and $i(\rho)$ represents the enthalpy function, and is usually a power law for the density ρ that satisfies $i(\rho)$, $i'(\rho) > 0$. The constant K, Bernoulli's constant, is required to be positive.

Existence of physically meaningful solutions to system (1.1), i.e., entropic weak solutions, remains an unsolved problem for any domain or boundary values. Hence, in an attempt to construct entropic solutions to system (1.1), viscous approximation models are usually considered, with the hope that they can be solved, and have sufficient uniform bounds in the viscosity parameter in order to provide compactness results that would yield the existence of entropic solutions to the inviscid system (1.1) by vanishing viscosity methods. A good approximation to transonic flow models (that is, steady potential flow) is given by

$$\text{div}(\rho \nabla \varphi) = 0, \tag{1.2}$$

$$\frac{1}{2}|\nabla \varphi|^2 - (K - i(\rho)) = \nu\, g(|\nabla \varphi|)\, \Delta \varphi. \tag{1.3}$$

For $\nu = 0$, (1.3) is just Bernoulli's Law. See as references for this viscous formulation and its justification, Courant and Friedrichs [15], Serrin [49], Synge [51], Morawetz [40]. Synge [51] showed exponential solutions for a linearized thermodynamical system of a viscous fluid that conducts heat, near constant equilibrium states, when specific entropy and specific volume are taken as basic thermodynamical variables. In particular, he stressed the difference in the constitutive form of the viscous terms when comparing hydrodynamics with thermodynamics.

Numerical experiments have been widely developed for system (1.1). For a survey on numerical simulations for transonic flow and approximations, see Jameson [32] and references therein.

In a recent work, see [25], we posed and solved a boundary value problem to a class of potential fluid-Poisson systems, which includes system (1.2)–(1.3). There we show the existence of smooth strong solutions that have uniform bounds in the viscosity parameter. More precisely, we show that there is a one-parameter family of solutions (φ^ν, ρ^ν) which are infinitely differentiable in the flow domain, with ρ^ν and $|\nabla \varphi^\nu|$ uniformly bounded in ν, and strictly positive for ν fixed.

This is a first and fundamental step in order to achieve a convergence in ν result that would yield a weak solution for the inviscid problem, (1.1) in standard gas dynamics or the larger problem for inviscid hydrodynamic-Poisson systems.

In the case of two-dimensional transonic flow, constructing weak solutions of system (1.1) from solutions of (1.2)–(1.3) has been outlined by Morawetz in [40], [41] using methods of compensated compactness presented by Murat [42], Tartar [52], and Di Perna [17], and G.Q.Chen [11] for the initial value problem for the one-dimensional time-dependent compressible fluid system of two equations, as in 1-dimensional isentropic gas dynamics with a power pressure law. Recently, 1-dimensional isentropic gas dynamics has been solved in unbounded domains by means of artificial viscosity and perturbations, for any positive initial density; see Lions, Perthame, and Tadmor [35] and Lions, Perthame, and Souganidis [36].

All these methods require uniform estimates in the parameters of the approximation. Clearly, if we pursue a weak solution (φ^0, ρ^0) of problem (1.1) which is a limit of some subsequence of solutions (φ^ν, ρ^ν) of (1.2)–(1.3), then it is to be expected, if there is strong convergence, that $i(\rho^0)$ and $\frac{1}{2}|\nabla\varphi^0|^2$ lie between 0 and K, and hence $i(\rho^\nu)$ and $\frac{1}{2}|\nabla\varphi^\nu|^2$ should also be between 0 and K up to, at most, an $\mathcal{O}(\nu^\beta)$-correction.

In fact, we have shown in [25] that, if the speed $|\nabla\varphi^\nu|$ is prescribed in a section of the boundary adjacent to two streamline boundaries, then normally $0 < k_\nu < i(\rho^\nu) < K$, but

$$0 < \frac{1}{2}|\nabla\varphi^\nu|^2 \leq \sup \frac{1}{2}|\nabla\varphi^\nu|^2 \leq \widetilde{K}, \tag{1.4}$$

where $\widetilde{K} = (K + C\nu^{1/2})M$, with M a constant that depends only on the flow domain and C depends on the boundary data and the flow domain.

Later, Gamba [24] showed that any smooth solution (ρ^ν, φ^ν) of (1.2)–(1.3), that satisfies estimate (1.4) (potential isentropic gas flow case), also satisfies

$$0 < \frac{1}{2}|\nabla\varphi^\nu(x)|^2 \leq K + C\,\frac{\nu}{(\operatorname{dist}\{x,\partial\Omega\})^2}, \tag{1.5}$$

for any x in the interior of the 2- or 3-dimensional flow domain Ω, and a growth condition on the enthalpy function $i(\rho)$, to be specified below. In particular, if the enthalpy function is the one associated with a γ-pressure law with $1 < \gamma$, then the necessary growth condition is satisfied for $1 < \gamma < 2$ in the 2-dimensional case and $1 < \gamma < \frac{3}{2}$ in the 3-dimensional one. The constant C depends on $\Omega, \widetilde{K}$, and the growth conditions for the functions $i(\rho)$ and $g(|\nabla\varphi(x)|)$ from (1.3).

In addition, for the two-dimensional case (where [25] showed existence of solutions for a boundary value problem associated with (1.2)–(1.3) that satisfied estimate (1.4)), we extend estimate (1.5) to some boundary points x in $\partial\Omega\setminus\partial_3\Omega$, where $\partial_3\Omega$ denotes the section of the boundary of the flow domain Ω where the speed was prescribed (named outer boundary). The parameter ν is replaced by ν^β for these boundary estimates, and the exponent β depends on the location of x in $\partial\Omega\setminus\partial_3\Omega$, and $\mathcal{C}$ denotes a number that depends on the local curvature of the boundary at the point x and the data of the boundary problem and the coarse bound $\widetilde{K}$. In fact $\mathcal{C}$ is bounded by a function of the Jacobian transformation that corresponds to the conformal map that takes Ω into a rectangle.

In addition, [24] gave an estimate similar to (1.5) for the potential fluid-Poisson system presented below. In this case the estimate reads

$$0 < \frac{1}{2}|\nabla\varphi^\nu(x)|^2 + \mathcal{R}(\varphi^\nu) - q\Phi^\nu| \leq K + \mathcal{C}\,\frac{\nu^{1/2}}{(\operatorname{dist}\{x,\partial\Omega\})^2}, \quad (1.6)$$

also for any x in the interior of the two- or three-dimensional flow domain Ω, and for the same growth condition on the enthalpy function $i(\rho)$ as in the gas flow case. Here the constant $\mathcal{C}$ depends on Ω, $\widetilde{K}$, the growth conditions for the functions $i(\rho)$ and $g(|\nabla\varphi(x)|)$, and the bounds on $\mathcal{R}$ and Φ_ν (there are proven to be ν independent bounds in the two-dimensional existence theory).

We point out that *cavitation speed* in isentropic gas flow is the constant value $(2K)^{1/2}$. However, for the hydrodynamic fluid-Poisson system, *cavitation speed* is not constant any longer. It is the speed at vacuum state given by the model, i.e. $|\nabla\varphi^\nu(x)|$ reaches *cavitation speed* when it takes the value $\{2(K - \mathcal{R}(\varphi^\nu(x)) + q\Phi^\nu)(x)\}^{1/2}$. Hence, the convergence analysis in the limiting vanishing parameter ν will also need estimate (1.6).

This estimate deteriorates as x is at $\nu^{\beta/2}$-distance from the outer boundary $\partial\Omega_3$, with $\beta = 1, \frac{1}{2}$, suggesting the possible formation of large boundary layers near $\partial\Omega_3$, as expected from viscous approximations in bounded domains. Thus, boundary layers yielding values of the speed that are larger than *cavitation speed* can **only** be formed at distances of order $\nu^{\beta/2}$ from the outer boundary.

It appears that the estimate (1.5) will still hold even if the section $\partial_3\Omega$ of the boundary is taken to infinity, as the conformal map that takes Ω into an infinite strip tends to the identity map at infinity. As a consequence, the solutions would have speeds that remain smooth and uniformly bounded in ν as the outer boundary section $\partial_3\Omega$ is absorbed at infinity.

In the following section, we present the potential fluid-Poisson model and previous results for the boundary value problem; subsequently, we outline the results already obtained. Then, we shall present an improvement for the lower bound of the density.

2 Presentation of the Problem in the General Case

We begin with the steady-state conservation laws for mass and momentum, and couple them to the Poisson equation for the electric field. The principal variables are charge density ρ, velocity $\vec{u}$, pressure P, and electric field $\nabla\Phi$, where Φ is the electric potential. Thus, with $x \in \mathbb{R}^n, n = 2$ or 3, the conservation laws are:

$$\operatorname{div}(\rho\vec{u}) = 0, \tag{2.1}$$

$$m\rho(\vec{u} \cdot \nabla)\vec{u} + \nabla P = q\rho\nabla \cdot \Phi + \vec{F}, \tag{2.2}$$

and the Poisson equation is

$$\epsilon\Delta\Phi = q(\rho - C(x)). \tag{2.3}$$

The parameters m, q, ϵ are, respectively, electron effective mass (parabolic band approximation, see [7]), space charge constant, and dielectric constant. The vector function $\vec{F}$ represents forces caused by viscosity or friction, and we chose $\vec{F}$ such as gave us a solvable set of equations that is as consistent with the physics as has proved possible, and is nowhere at great variance with the physical problem. $C(x)$ is the doping profile function and represents the background charge. It is assumed that $C(x)$ is a non-negative step function.

In the approximation to the transonic problem, $\Phi \equiv 0$ and (2.3) does not appear. The conservation of energy is replaced by taking the pressure P as a given function of $m\rho$, the mass density. This is consistent with the notion that the disturbances we are studying are weak. For example, and for simplicity, suppose we consider the fluid dynamical case with $\vec{F} = 0$ and now suppose $P = P(\rho, S)$, where S is entropy. The full conservation laws then admit shocks, but a change in entropy across a shock would be third order in the strength of the shock. Thus, we may take $P = P(\rho)$, provided third-order errors may be neglected.

2.1 Potential Flow and the Choice of the Viscous–Friction Force Term $\vec{F}$

If $\vec{F} \equiv 0$, we could look for an irrotational flow, $\operatorname{curl} \vec{u} = 0$, so that $\vec{u} = \nabla\varphi$, where φ is the potential, and reduce the system to conservation of mass, a Bernoulli Law, and Poisson's equation. An irrotational flow is again consistent with the neglect of third-order terms. What is more, it is mathematically useful because the system is reduced to fourth-order (second-order if $\Phi \equiv 0$ and weakly coupled if $\Phi \neq 0$). Henceforth we assume

$$\vec{u} = \nabla\varphi, \tag{2.4}$$

and choose $\vec{F}$ so that we can make a similar reduction in order. Then we check the physical consistency of this choice of $\vec{F}$. Thus, a preliminary choice would be:

$$\vec{F} = \nabla\Psi,$$
$$\Psi = \nu\Delta\varphi - K\varphi,$$

where ν is a coefficient of viscosity and K a coefficient of friction. Then, with $\vec{u} = \nabla\varphi$, the momentum equation (2.2) can be integrated to yield Bernoulli's law:

$$\frac{1}{2}m|\nabla\varphi|^2 + i(\rho) = -q\Phi + \nu\Delta\varphi + K\varphi + \text{const.}$$

Here $i(\rho)$ is the enthalpy, i.e., $i(\rho) = \int \frac{dP(\rho)}{\rho}$, where

$$\begin{cases} i(\rho) \geq 0, & \text{if } \rho \geq 0, \\ i(\rho) = 0, & \text{otherwise}, \end{cases} \tag{2.5}$$

and $i'(\rho) > 0$ if $\rho > 0$. Furthermore we assume (b) $P(\rho)$ is convex, or $i''(\rho), i'''(\rho)$ are nonnegative.

We assume in this chapter that the constant K is independent of ν, and is adjusted by the data in order to be consistent with the inviscid problem (i.e., $\nu = 0$).

The first term in $\vec{F}$ is $\nu\rho\Delta\vec{u}$, which is the appropriate viscous force, and the second is $\tau_p^{-1}m\rho\vec{u}$, which represents a friction term. The viscous term, with ν a constant coefficient of viscosity, is essentially the same as that of Serrin [49] or Synge [51]. The friction term is, as in Baccarani and Wordeman [4] and Odeh, Gnudi, and Rudan [30], the momentum density divided by a constant velocity relaxation time τ_p.

However, we cannot solve the b.v.p. with this choice of $\vec{F}$. To solve the system, the viscous term in Ψ, i.e., $\nu\Delta\varphi$, must be modified to approach infinity at zero speed like $|\nabla\varphi|^2$, and to approach zero at infinite speed like $|\nabla\varphi|^p$, $p \geq 3$. For similar reasons, τ_p must become infinite as $\varphi \to \pm\infty$. Thus, the force term $\vec{F}$ is given by

$$\vec{F} = \nabla\Psi, \quad \Psi = \nu\mathcal{G}\Delta\varphi - \mathcal{R}(\varphi), \tag{2.6}$$

where $\mathcal{G}$ and $\mathcal{R}$ have the desired properties.

The case of interest is low viscosity, ν small. It turns out that $|\nabla\varphi|$ and φ are uniformly bounded in our b.v.p. In the range of speed and potential that occur, $\mathcal{G}(|\nabla\varphi|) \sim 1$ (except near zero speed) and $\mathcal{R}(\varphi)$ is linear. Thus, the restrictions on viscous and friction coefficients have little effect on the physical interpretation of the result.

Therefore, using (2.6) for the force term, we obtain Bernoulli's Law as follows:

$$\frac{m}{2}|\nabla\varphi|^2 + i(\rho) = K + q\Phi - \mathcal{R}(\varphi) + \nu\mathcal{G}\Delta\varphi. \tag{2.7}$$

The conservation of mass (2.1) becomes for potential flow,

$$\text{div}\rho\nabla\varphi = 0. \tag{2.8}$$

For the two-dimensional case, it is also possible to define a stream function ψ from (2.1), whose level curves describe the particle path at any point and are orthogonal to the corresponding level curves of the potential flow φ, i.e., with $\rho\vec{u} = (-\psi_y, \psi_x)$,

$$\text{div}(\tau\nabla\psi) = 0, \tag{2.9}$$

where $\tau = \rho^{-1}$.

That motivated us to seek a boundary value problem for which we can prove existence, and also establish the conditions that might eventually lead to the existence of a corresponding limiting solution as the parameter ν becomes very small with respect to the other constants involved in the model, i.e. the existence of a corresponding limiting inviscid solution as proposed by Morawetz [40] in order to solve the transonic gas flow model. Such a program (see [41]) cannot be carried out without better bounds than we get for the state variables, ρ and $\vec{u} = \nabla\varphi$.

In [25], we focussed on the existence of solutions for the "viscous" potential fluid-Poisson system (2.7)–(2.9) and (2.3) in "smooth" four-sided domains and with boundary conditions relevant to the physical problem: two opposite tangential flow walls, an inflow condition determined by prescribing a constant potential flow function, and prescribed speed in the remaining part of the boundary, which we shall call the outer boundary. These boundary conditions are to be described in detail later; nevertheless we add here that in the case of gas flow past a profile, our outer boundary condition means that "almost" Mach number speed is prescribed if ν is very small.

Also, as in the case of approximations to isentropic flow past a profile, we proved existence of a regular solution such that the density and speed are uniformly bounded in ν, away from zero and infinity, and the electric field is also uniformly bounded in ν, with bounds similar to those in (1.4) and (1.5).

A physically more correct one-dimensional system (2.1)–(2.3) was solved in [20]. There, it is proved that the solution triple ρ, v and Φ is of uniformly bounded variation, and that there exists a limiting inviscid solution ρ^0, v^0, and Φ^0 of bounded variation, such that the convergence from the viscous solution to the inviscid one is pointwise and in $L^1(\Omega)$. In addition, ρ^0 and v^0 might have admissible shocks, that is, discontinuities that keep the mass and momentum flux terms Lipschitz continuous functions that satisfy the "entropy condition", i.e., the density increases across the discontinuity in the direction of the particle path. In addition, the possible formation of boundary layers was analyzed in [20] and [22].

Remark Setting $\nu = 0$ in (2.7), we obtain a transonic type of equation for the inviscid model. Indeed, taking $i(\rho) = K + q\Phi - \mathcal{R}(\varphi) - \frac{m}{2}|\nabla\varphi|^2$, defining $c^2 = i'(\rho)\rho = \frac{dP}{d\rho}$ as the *local speed of sound*, and combining with the flow equation (2.6), we find that the potential flow function φ satisfies the p.d.e.:

$$(c^2 - m\varphi_x^2)\varphi_{xx} - 2m\varphi_x\varphi_y\varphi_{xy} + (c^2 - m\varphi_y^2)\varphi_{yy} = S(x, y, \varphi, \varphi_x, \varphi_y, \rho, \Phi, \nabla\Phi),$$

which is a **mixed-type** equation for φ, i.e., an elliptic equation for values of the speed below $c m^{-1/2}$, and hyperbolic for values of the speed above $c m^{-1/2}$.

2.2 Further Conditions

The function $\mathcal{R}$ is a Lipschitz function of φ and $\mathcal{R}(\varphi)$ is bounded above and below by constants $\mathcal{R}_U$ and $\mathcal{R}_L$, respectively.

The function $\mathcal{G}$ of the "viscous" term $\nu \mathcal{G} \Delta \varphi$ is taken as $\mathcal{G} = G(|\nabla \varphi|)$, bounded below by a positive constant C_1, where G and G^{-1} are locally Lipschitz functions and satisfy

$$t^2 G(t) \to C_1 > 0, \quad \text{as} \quad t \to 0, \tag{2.10}$$

and

$$t^{-1} G(t) \to C_2 > 0, \quad \text{as } t \to \infty, \text{ and } t^2 \frac{G'(t)}{G^2(t)} \text{ is bounded.} \tag{2.11}$$

The enthalpy function $i(\rho)$ satisfies $i(\rho), i'(\rho) > 0$, and the growth condition,

$$\frac{1}{n-1} i(\rho) - \rho\, i'(\rho) \geq \frac{1}{n-1} k\, i(\rho), \quad \text{for some} \quad k,\ 0 < k < 1, \tag{2.12}$$

where $n = 2$ or 3 denotes the space dimension.

For $\mathcal{R}_L < \mathcal{R}(\varphi) < \mathcal{R}_U$, the Bernoulli constant K satisfies the compatibility condition,

$$K - \mathcal{R}_U + q\Phi_L(K) > 0, \tag{2.13}$$

where

$$\Phi_L(K) = \inf_{\partial_1 \Omega \cup \partial_3 \Omega} \gamma - \alpha \sup_{\Omega} |F_x|^2 (\sup_{\Omega} C(x) + i^{-1}(K - \mathcal{R}_L + q\Phi_U)),$$

for $\Phi_L(K) \leq \Phi^\nu \leq \Phi_U$, where $\Phi_L(K, \Phi_U)$ are ν-independent, and depend on the domain Ω and the data of the problem.

Remark Equations (2.1)–(2.2) correspond to a dissipative approximation to a compressible flow model that satisfies the γ-law: $i(\rho) = \frac{\gamma}{\gamma-1} \rho^{\gamma-1}$, so condition (2.3) is satisfied with $k = 2 - \gamma$ and $1 < \gamma < 2$, if the space dimension satisfies $n = 2$; and $k = 3 - 2\gamma$ and $1 < \gamma < \frac{3}{2}$ if $n = 3$.

We point out here that conditions (2.10)–(2.11) are related to the existence problem as well as the problem of getting uniform bounds, as has been shown in [25]. However, it is quite interesting to remark that a condition of the same kind has been used in computational transonics in the construction of *preconditioner matrices* proportional to the Mach number for computational compressible viscous flows with very high Reynolds numbers in regions of low Mach flow (see Choi and Merkle [13] and Nigro, Storti, Idelsohn, and Tezduyar [44]).

However, since we prove that $|\nabla \varphi^\nu|$ is uniformly bounded above by a number M that depends only on the domain and the data of the boundary value problem for any $\nu \leq \nu_0$, with ν_0 depending also on the domain and the data of the problem, a posteriori we get an existence result and upper uniform bounds even if we take a viscosity that is linear away from stagnation points, i.e., by setting

$$G(|\nabla\varphi|) = 1 \quad \text{for } 0 < k_\nu < |\nabla\varphi| \le M^{\frac{1}{2}},$$

and extending it smoothly to $[0, \infty]$, such that growth conditions (2.10) and (2.11) are satisfied. This is desirable for physical reasons.

Thus, we shall assume here that g is a monotone increasing function and

$$g(|\nabla\varphi|) = \left(1 + \frac{|\nabla\varphi|^2}{2}\right)^{\alpha/2}, \quad \alpha \ge 1, \tag{2.14}$$

for sufficiently large values of $\frac{|\nabla\varphi|^2}{2}$.

We point out that the compatibility condition (2.13) reduces to $K > 0$ for the approximation to the transonic flow model, which is the standard assumption on Bernoulli's constant. In the case of the fluid-Poisson system under consideration, since $i^{-1}(\rho)$ has superlinear growth, then a K, verifying (2.4), exists only if the data are chosen adequately.

By equation (2.8), the introduction of $\frac{\partial}{\partial\varphi}$, as a derivative along the streamline, yields

$$\Delta\varphi = -\frac{\nabla\rho}{\rho} \cdot \nabla\varphi = -(\ln\rho)_\varphi |\nabla\varphi|^2. \tag{2.15}$$

Using (2.12) in (2.7), we obtain from (2.3) and (2.6) the complete system of equations for φ, ρ, and Φ. We add the equation in ψ for convenience:

$$\operatorname{div}(\rho\nabla\varphi) = 0, \quad \operatorname{div}(\tau\nabla\psi) = 0, \tag{2.16}$$

$$\frac{m}{2}|\nabla\varphi|^2 + i(\rho) + \mathcal{R}(\varphi) - q\Phi - K + \nu(\ln\rho)_\varphi |\nabla\varphi|^2 \mathcal{G} = 0, \tag{2.17}$$

$$\Delta\Phi = \alpha(-C(x) + \rho), \quad \alpha = \frac{q}{\varepsilon}. \tag{2.18}$$

Here $\tau = \rho^{-1}$ represents specific volume. Note that equation (2.17) states that the density is governed by the history of the flow on the streamline (analogous to upwind differentiation in some sense). Indeed, heuristically, equation (2.17) is the result of the first order expansion of the density values corresponding to "retarded" flow speed, along the streamline, with an order of magnitude

$$\nu \frac{G(|\nabla\varphi|)\,|\nabla\varphi|}{(\gamma-1)\,i(\rho)}(x_r) = \nu M^2 \frac{G(|\nabla\varphi|)}{|\nabla\varphi|}(x_r), \tag{2.19}$$

with $M(x_r) = \frac{|\nabla\varphi|}{(\gamma-1)\,i(\rho)}(x_r)$ the local Mach number. Thus, equation (2.17) is the result of the first-order formal Taylor expansion of the density ρ for the equation

$$\frac{|\nabla\varphi(x_r)|^2}{2} + i\left(\rho\left(x_r + \nu\frac{G(|\nabla\varphi|)}{(\gamma-1)\,i(\rho)}\nabla\varphi(x_r)\right)\right) + \mathcal{R}(\varphi)(x_r) - q\Phi(x_r) = K. \tag{2.20}$$

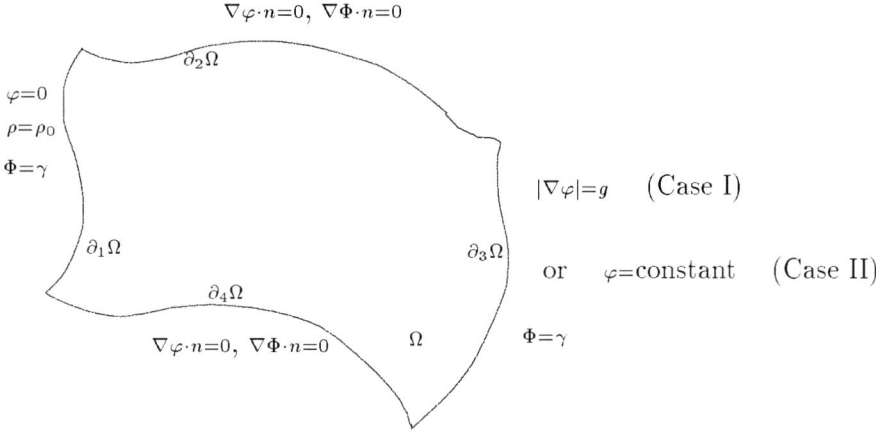

FIG. 1. The flow domain Ω and boundary data

Remark System (2.16)–(2.18) reduces to a two-by-two system, in the case of the gas flow problem, with a second-and first-order equation. Conditions (2.10)–(2.11) modified the time-like step for the first-order equation by a factor proportional to the Mach number and speed. These conditions are necessary for the existence problem, as well as for obtaining uniform bounds with respect to the viscous parameter, as has been shown in [25]. However, we remark that a condition of the same kind as (2.10) has been used in computational transonics in the construction of *time–derivative preconditioner* algorithms for computational compressible viscous flows with very high Reynolds numbers in regions of low Mach flow. The preconditioner matrices are proportional to the Mach number and density as well (see Choi and Merkle [13] and Nigro, Storti, Idelsohn, and Tezduyar [44]).

2.3 The Domain and Boundary Conditions

We take into consideration two possible boundary value problems in the case of the two-dimensional flow model. One of them is the model with data given in [25]; we prescribe there an inflow boundary, two adjacent tangential flow boundaries (i.e., two walls), and in the remainder, we prescribe positive non-cavitating speed (i.e., the magnitude of the velocity field).

The other boundary value problem is one that corresponds to prescribing an outflow boundary condition (i.e., the flow potential φ is a positive constant) on the section where the speed was prescribed in the above case.

In fact, we consider some special 2-D flow domains: Let Ω be as in Figure 1 above. Thus, the boundary of Ω is the union of four smooth curves that meet each other at a right angle. In particular, there is a unique conformal transformation that takes Ω into a rectangle R which keeps three points fixed (take any three of the angle points, including the two that correspond to the inflow boundary meeting the tangential flow segments). In addition, the conformal map is smooth (C^3). We denote the boundary sections as follows: $\partial \Omega_1$ the inflow boundary section, $\partial \Omega_2$ and $\partial \Omega_4$ the tangential flow boundary sections, and $\partial \Omega_3$ the remainder portion.

Hence, the two boundary value problems under consideration both have the same data on the inflow and tangential flow boundary sections, viz., the potential flow function $\varphi = $ constant on $\partial_1 \Omega$, with $(\nabla \varphi \cdot n|_{\partial_1 \Omega})(\omega_1) < 0$ for ω_1 a corner point where $\partial_1 \Omega$ meets $\partial_2 \Omega$; $\nabla \varphi \cdot n = 0$ on $\partial_\tau \Omega = \partial_2 \Omega \cup \partial_4 \Omega$. As usual, n denotes the outer unit normal.

The density is prescribed at the inflow boundary, i.e., at $\partial_1 \Omega$, so that $\rho = r(x)$ on $\partial_1 \Omega$ and the electric potential Φ satisfies Dirichlet conditions, i.e., $\Phi = \gamma$ on $\partial_1 \Omega \cup \partial_3 \Omega$ and $\nabla \Phi \cdot n = 0$ on $\partial_\tau \Omega$.

Thus, the first boundary value problem prescribes $|\nabla \varphi| = g(x) > 0$, on $\partial_3 \Omega$, and the second just $\varphi = $ constant on $\partial_3 \Omega$ (i.e., an outflow boundary if this constant is larger than the one for the inflow boundary).

2.4 Existence and Uniform Bounds

In [21] and [25] we solved the first boundary value problem for a more general system of equations than (2.13), viz.,

$$\mathrm{div}(\rho \nabla \varphi) = 0, \tag{2.21}$$
$$-\nu(\ln \rho)_\varphi = f(|\nabla \varphi|^2, \theta, Q_B(\rho, \varphi, \Phi)), \tag{2.22}$$
$$\Delta \Phi = \alpha(\rho - C(x)), \tag{2.23}$$

where Q_B denotes the squared speed given by Bernoulli's equation (i.e., equation (2.6) for $\nu = 0$) and $\theta = \arctan \frac{\varphi_y}{\varphi_x}$, the directional angle of $\nabla \varphi$, taken to be zero at some point on the inflow boundary.

We showed [25] that solution triples ρ, φ and Φ, in $C^{1,\alpha}$, $C^{2,\alpha}$ and $W^{2,p}$ respectively, exist for sufficiently small $\nu > 0$ under conditions (2.10)–(2.11). In addition, $0 < i(\rho) \leq L^*$ for all ν, where L^* depends only on the Bernoulli constant K, the bounds for the function $\mathcal{R}$, the domain Ω, and the data of our boundary value problem (the latter controls $\sup \Phi^\nu$).

Then, Q_B takes the value $Q_B = K - i(\rho) - R(\varphi) + q\Phi$, which is the square of the speed given by Bernoulli's Law, as long as $Q_B > 0$ (see [25]). Thus, any solution of system (2.21)–(2.23), with

$$f(|\nabla \varphi|^2, Q_B) = \left(\frac{m}{2}|\nabla \varphi|^2 - Q_B\right)\left(|\nabla \varphi|^2 G(|\nabla \varphi|)\right)^{-1}, \tag{2.24}$$

yields a solution of system (2.16)–(2.18).

In addition, uniform bounds for φ and $|\nabla \varphi|$ hold for the particular case $\nu \ll m$ in an appropriate length scale. That is, if $i(\rho)$ satisfies condition (2.12)

and $\mathcal{G}$ satisfies (2.10)–(2.11), there is a $\nu_0 = \nu_0(\mathcal{G}, \Omega)$ such that
$$0 < k_\nu < |\nabla \varphi^\nu| \le K^*, \quad 0 < \ell_\nu < \rho^\nu < L^*, \quad \text{for} \quad \nu \le \nu_0, \tag{2.25}$$
and $|\nabla \Phi^\nu|, |\Phi^\nu| \le M^*$ with bounds all taken in $\overline{\Omega}$, with K^*, L^* and M^* independent of ν. In fact, the upper bound $\widetilde{K}$ for the speed is given in terms of cavitation speed and the domain Ω. That is, if $F \colon \Omega \to R$ is the conformal map that takes Ω into a rectangle R, then
$$K^* = \{\sup_{\overline{\Omega}} |K - \mathcal{R}(\varphi) + q\Phi| + C\nu^{1/2}\} \frac{\sup_{\overline{\Omega}} |F_x|}{\inf_{\overline{\Omega}} |F_x|}, \quad \nu \le \nu_0, \tag{2.26}$$
where $|F_x|$ is the Jacobian of the real-valued transformation associated with F, $\nu_0 = \nu_0(k, \mathcal{G}, \|F\|_{C^{1,1}(\overline{\Omega})})$ and $C = C(k, \mathcal{G}, \|F\|_{C^{1,1}(\overline{\Omega})}, K, q, \|\mathcal{R}\|_{C^{0,1}(\overline{\Omega})}$, boundary data).

As we stated above, estimate (2.25) is not sharp, since it predicts that the speed corresponding to the viscous flow is bounded away from a factor of *cavitation speed* by a term of order $\mathcal{O}(\nu^{1/2})$. Indeed, the factor is
$$\sup_{\overline{\Omega}} |F_x| \cdot \{\inf_{\overline{\Omega}} |F_x|\}^{-1} = \exp\{\operatorname{osc}(\log |F_x|)\},$$
and it is related to the geometry of the domain, a sort of measure of how far the flow domain is from a rectangle, since
$$\operatorname{osc}(\log |F_x|) \le \sup_{\Omega} |\nabla (\log |F_x|)| \operatorname{diam}(\overline{\Omega}) \le \sup_{\overline{\Omega}} |D_{ij} F| (\inf |F_x|)^{-1} \operatorname{diam}(\overline{\Omega}).$$

Unless the domain Ω is originally a rectangle (or ν-close to a rectangle), the value of K^* is too coarse.

The proof of the sharper estimate requires the existence of an approximate φ^ν solution to (2.16)–(2.18) under conditions (2.10)–(2.13), such that $|\nabla \varphi^\nu| < K^*$, where K^* is a ν-uniform constant that depends on the data and the flow domain Ω (see [25] for the two-dimensional case).

Remark The existence of three-dimensional solutions to system (2.1)–(2.2) is an open problem. However, under the assumption of existence of solutions $(\rho^\nu, \varphi^\nu, \Phi^\nu)$ to a three-dimensional boundary value problem, where conditions (2.10)–(2.13) are satisfied, and $|\nabla \varphi^\nu|$ admits a ν-uniform bound denoted by $\widetilde{K}$, then $|\nabla \varphi^\nu|$ satisfies a sharper pointwise bound "close" to cavitation speed in the interior of the flow domain.

Hence, the following pointwise estimate was proven for $|\nabla \varphi^\nu(x)|$ in the interior of the two- or three-dimensional domain, for any solution of a boundary value problem associated with system (2.10)–(2.13)/(2.16)–(2.18), where estimate (2.6) is satisfied ([24]):

$$|\nabla \varphi^\nu(x_0)|^2 \le 2(K - \mathcal{R}(\varphi^\nu(x_0)) + q\Phi^\nu(x_0)) + \frac{C\nu^\beta}{(\operatorname{dist}(x_0, \partial\Omega))^2}, \quad \text{for all} \quad \nu \le \nu_1, \tag{2.27}$$

where ν_1 and C both depend on $k^{-1}, K^*, K, g, q, \alpha$, the bounds of $\mathcal{R}$ and Φ. The parameter ν_1 depends on dist$\{x_0, \partial\Omega\}$. The exponent β is 1 for the compressible gas model, and $1/2$ for the fluid-Poisson system, but it is independent of the space dimension.

Furthermore, in the two-dimensional case, we can extend the pointwise estimate to the boundary region, which includes the inflow and tangential flow regions. Therefore, for $\Omega \subset \mathbb{R}^2$,

$$|\nabla\varphi^\nu(x_0)|^2 \leq 2(K - \mathcal{R}(\varphi^\nu(x_0)) + q\Phi^\nu(x_0)) + \frac{C\nu^\beta}{(\text{dist}(x_0, \partial_3\Omega))^2}, \quad \text{for all} \quad \nu \leq \nu_1, \tag{2.28}$$

where $\beta = \frac{1}{2}$ if the point x_0 lies on the tangential boundary $\partial_\tau \Omega = \partial_2\Omega \cup \partial_4\Omega$. On the other hand, $\beta = \frac{1}{8}$ if x_0 lies on the inflow boundary $\partial_1\Omega$, and $\beta = 1$ if $\partial_1\Omega$ is locally flat around x_0 (i.e., $\kappa_x = 0$ in a neighborhood of x_0 relative to $\partial_1\Omega$). In all cases, the constants ν_1 and C depend on $k^{-1}, g, K^*, K, \mathcal{R}, q, \alpha$, data, κ_{x_0}, with κ_{x_0} the local curvature of $\partial\Omega$ at x_0.

Clearly, estimate (2.28) allows the possible formation of large boundary layers near the outflow boundary $\partial_3\Omega$ at distances less than $\mathcal{O}(\nu^{1/2})$ away from the tangential boundary, and of order $\mathcal{O}(\nu^{1/4})$ near the tangential boundary. This behavior excludes velocity overshoots above cavitation speed for the viscous solutions near shock formation for the limiting configuration, away from $\partial_3\Omega$.

We remark that an interesting aspect of this technique is that it works in three dimensions as well. Provided there is available the existence of solutions with ν-independent bounds up to the boundary, these estimates seem not to depend on the conformal map, but rather on the local parameterization of the boundary to a flat one. As was described above, we do not include here boundary estimates in the three-dimensional case due to lack of knowledge of a boundary value problem that yields existence of solutions with a coarse ν-uniform bound for the speed. However, estimate (2.27) holds *under the assumption* of existence of solutions with ν-independent bounds for the speed up to the boundary of the flow domain.

Finally, in the two-dimensional case, the following conclusion holds. Assume existence of smooth solutions for the second boundary value problem presented above (case II), where φ^ν is prescribed on $\partial_3\Omega$ as an arbitrary constant above the cavitation ratio (i.e., the ratio between cavitation speed and the length of the shortest curve among those that define the tangential flow walls for the domain Ω). Assume conditions (2.10)–(2.13) (i.e., $\partial_3\Omega$ is now an outflow boundary). Then, $\varphi^\nu \in C^{2,\alpha}(\overline{\Omega})$ and $0 < k_\nu \leq |\nabla\varphi^\nu| < K_\nu$, but $|\nabla\varphi^\nu|$ cannot be ν-uniformly bounded.

Finally, we look at the lower estimate ℓ_ν for the density, from (2.25). It was shown to be of order $e^{-\frac{1}{\nu}}$ (see (3.24)–(3.25) in [25]). As $\nu \to 0$, this very rapid decay can be considerably improved for the case of gas flow past a profile.

Theorem 2.1. (Gas flow past a profile case) *Let ρ^ν be the $C^{1,\alpha}$ solution* of (1.2)–(1.3) constructed in [25] where the solvability conditions are satisfied. If, in addition to (2.10)–(2.11), the function $G(t)$ satisfies $0 < C_1 \leq t^2 G(t)$ for $t > 0$, then

$$C\nu \leq i(\rho^\nu)(x) \leq L^*, \quad \text{for any } x \in \overline{\Omega} \setminus \mathcal{D}_\nu, \tag{2.29}$$

where $C = C(\Omega, \gamma, C_1, K^*, L^*, d^{-2})$ for $d = \text{dist}(x, \partial_3 \Omega)$, independent of ν for $\nu \leq \nu_0(\mathcal{G}, \Omega)$. The set $\mathcal{D}_\nu$ satisfies $\partial \Omega \subset \mathcal{D}_\nu \subset \overline{\Omega}$ and contains a "slit" of the outer boundary $\partial_3 \Omega$ such that $\text{dist}\{\overline{\Omega} \setminus \mathcal{D}_\nu, \partial_3 \Omega\} = \mathcal{O}(\nu^{1/2 - \delta})$, $\delta > 0$.

Theorem 2.2. (Charged-particle system case) Let ρ^ν be the $C^{1,\alpha}$ solution of (2.16)–(2.18) constructed in [25] where the solvability conditions are satisfied. If, in addition to (2.10)–(2.13), the function $G(t)$ satisfies $0 < C_1 \leq t^2 G(t)$ for $t > 0$, then

$$C\nu^{1/2} d^{-2} \exp(-\nu^{-1/2} \frac{\gamma - 1}{C_1} d^{-2}) \leq i(\rho^\nu)(x) \leq L^*, \quad \text{for any } x \in \overline{\Omega} \setminus \mathcal{D}_\nu, \tag{2.30}$$

where $C = C(\Omega, \gamma, C_1, K^*, L^*, d^{-2})$ for $d = \text{dist}(x, \partial_3 \Omega)$, independent of ν for $\nu \leq \nu_0(\mathcal{G}, \Omega)$. The set $\mathcal{D}_\nu$ satisfies $\partial \Omega \subset \mathcal{D}_\nu \subset \overline{\Omega}$ and contains a "slit" of the outer boundary $\partial_3 \Omega$ such that $\text{dist}\{\overline{\Omega} \setminus \mathcal{D}_\nu, \partial_3 \Omega\} = \mathcal{O}(\nu^{1/4 - \delta})$, $\delta > 0$.

Proof The proofs of both theorems are rather similar, so we shall write it once, and make the distinctions when needed. Let φ^ν, ρ^ν, and Φ^ν be the strong solutions constructed in [25]. Omitting the subscript ν, we find they satisfy the equation

$$-\frac{\nu}{\gamma - 1} \nabla \ln i(\rho) \frac{\nabla \varphi}{|\nabla \varphi|^2} = \frac{i(\rho) - \left(K - \frac{|\nabla \varphi|^2}{2} - \mathcal{R}(\varphi) + q\Phi\right)}{|\nabla \varphi|^2 G(|\nabla \varphi|)}, \tag{2.31}$$

because $i(\rho) = \frac{\gamma}{\gamma - 1} k \rho^{\gamma - 1}$.

Now, from estimate (2.28),

$$\left\{\frac{|\nabla \varphi|^2}{2} - (K - \mathcal{R}(\varphi) + q\Phi)\right\}(x) \leq C \frac{\nu^\beta}{d^2}, \quad \text{with } d = \text{dist}(x, \partial_3 \Omega), \tag{2.32}$$

for any $x \in \overline{\Omega} \setminus \partial_3 \Omega$, and $C = C(k^{-1}, G, K^*, K, \mathcal{R}, q, data, \kappa_x)$.

In addition, if $|\nabla \varphi|^2 G(|\nabla \varphi|) > C_1$ for all values $|\nabla \varphi|$, then the right hand side of (2.31) can be estimated, yielding the following differential inequality:

$$-\nu (\ln i(\rho))_\varphi \leq \frac{\gamma - 1}{C_1} \left(C \frac{\nu^\beta}{d^2} + i(\rho)\right), \tag{2.33}$$

for any $x \in \overline{\Omega} \setminus \partial_3 \Omega$.

Let $h = i(\rho) > 0$, $M = C \nu^\beta d^{-2}$ and $a = \frac{\gamma - 1}{C_1}$. Then (2.33) can be written as

$$-\frac{1}{a}(\ln h)_\varphi \leq M + h,$$

or equivalently,
$$-\frac{h_\varphi}{ah(M+h)} \leq 1.$$

Recalling integration by simple fractions, and integrating along the streamlines in the direction of growing φ, starting at $\Gamma \subset \partial\Omega$, we arrive at

$$\ln\left(\frac{M}{h}-1\right) \leq aM(\varphi - \varphi_0). \tag{2.34}$$

Hence, since $|\varphi| \leq \mathcal{K}(K^*)$ in $\overline{\Omega}$ uniformly in ν, it follows from (2.34) that

$$\ln\left(\frac{M}{h}-1\right) \leq aM\mathcal{K},$$

or, equivalently,

$$\frac{1}{a\mathcal{K}}e^{-(aM\mathcal{K})} \leq h. \tag{2.35}$$

Since
$$aM = \nu^{(\beta-1)}\frac{\gamma-1}{C_1}d^{-2},$$

we have obtained the following estimate from below for $i(\rho)$, viz.,

$$\frac{\nu C_1}{(\gamma-1)\mathcal{K}}e^{-(\nu^{\beta-1}d^{-2}\mathcal{K}\frac{\gamma-1}{C_1})} \leq i(\rho). \tag{2.36}$$

Therefore, for the gas flow past a profile model, $\beta = 1$ for all the points in the interior of the flow domain, so that with the selection of a "slit" next to the outer boundary of length $\nu^{1/2}$, estimate (2.29) follows with $C = \frac{C_1}{(\gamma-1)\mathcal{K}}e^{-(d^{-2}\mathcal{K}\frac{\gamma-1}{C_1})}$. Theorem 2.1 is then proved.

For the charged-particle transport model, the best we can get is $\beta = 1/2$ in the interior of the flow domain, which yields estimate (2.30) when the choice $\beta = 1/2$ is made in (2.36). This completes the proof of both cases. $\square$

2.5 Conclusions about Transonic Flow

Because $|\nabla \varphi^\nu|$ and ρ^ν are uniformly bounded from above, there exists a convergent subsequence (φ^ν, ψ^ν) with a limit (φ^0, ψ^0). However, this is not enough to establish that there exists a weak solution to the equations $\rho \varphi_x = \psi_y$, $\rho \varphi_y = -\psi_x$.

One method to prove the existence of weak solutions involves the method of compensated compactness (see Murat [42], Tartar [52], and Di Perna [17]). In the applications to mixed-type systems (see Morawetz [40], [41]), this has been used mainly for hyperbolic problems. This requires that the speed $|\nabla \varphi^\nu|$ is uniformly bounded below from zero and above from the cavitation speed (see [41]). Furthermore, one also requires bounds on the flow angle.

While this paper thus provides a complete proof of existence of viscous solutions with some uniform bounds which allows us to consider a convergent subsequence, still there is a major gap in showing that its limit is a solution of the inviscid problem.

Similar difficulties arise in the existence theorems of Feistauer and Nečas [19]. They show the existence of a solution for the inviscid model under the the assumption of existence of viscous solutions to a boundary value problem, provided the divergence of the viscous velocity field satisfies uniform bounds in ν. Gittel [29] shows existence using a variational approach to a boundary value problem for the transonic small disturbance equation and shows existence theorems assuming a priori uniform bounds and entropy conditions.

We also mention the work of Klouček and Nečas [43] and [34] to find entropic solutions for the transonic flow model by the method of stabilization. There they solve a perturbed flow equation by introducing an artificial time variable. Also there, they need to assume uniform speed and entropy bounds in order to pass to the limit.

2.6 About the Semiconductor Device Model

We have shown the existence of a regular solution to the boundary value problem for the fluid-Poisson system in an approximating geometry to a real device, and without smallness assumptions on the size of the data. The model we have considered corresponds to a fluid level approximation of a kinetic formulation for a charged particle system, with a pressure law similar to the one for an isentropic gas with $1 < \gamma < 2$, and a viscous parameter ν. In this context the "viscosity" parameter is related to the mean free path, and refers to the constant coefficient related to the the nonconvective energy flux term, which has been modeled as a nonlinear term that involves first-and second-order derivatives of the velocity field.

In addition, we show sharp uniform bounds in the viscosity parameter for the speed, and an improvement for the lower bound for the density, both valid for regimes where this parameter becomes small compared with the scaled coefficient of the acceleration term in the momentum equation. This is a potential flow model obtained under the assumption that the velocity field aligns with the gradient field of τ, where $\tau > 0$ stands for the velocity relaxation time term. Potential flow can be justified if the initial state has zero vorticity.

In order to extend our results to the case $\gamma = 1$ (the isothermal model), modifications would be required, especially to obtain the uniform bounds. This case is however often used in modeling: see [37] and [48].

In particular, for two-dimensional models in MOSFET geometry, it can be shown that the behavior of any solution at the boundary points will depend on the behavior and regularity at the boundary of the domain under the conformal transformation that takes the domain of the device into a rectangular domain, where the source and drain contacts are transformed into opposite walls of the transformed domain. Therefore, a singularity is expected to be found in the electric field at the boundary points corresponding to rough boundaries and to junctions between contact and oxide regions. In these cases, the hydrodynamic model worked out here for charged-particle transport cannot handle these boundary singularities for the electric field unless very restricted assumptions are given on the data. Then becomes necessary to look at models derived for transport regimes under strong force effects (see Barenger and Wilkins [5, 6], Poupaud [45, 46], Stichel and Strothmann [50], and Cercignani, Gamba, and Levermore [9, 10], for some mathematical work related to these models).

3 The Semiconductor

There are many models for semiconductors. They include the quantum, kinetic, and fluid level formulations. The most accurate is the quantum one, where the particles are represented by wave functions solving the Schrödinger equation with a Hamiltonian that incorporates potentials due to the semiconductor lattice, Coulomb interactions, the applied bias, and particle–phonon interactions. Although such models have begun to be used for numerical simulations of small parts of the semiconductor devices where quantum effects are important, up until now it has been impossible to formulate a model for a complete device at this level.

Then there is the kinetic level of modeling and the application of Monte-Carlo methods at the particle level, where the dynamics of the particles are described by the evolution of distribution functions for electrons and holes respectively, both depending on time, position, and wave-vector, where the latter belongs to a periodic lattice in $\mathbb{R}^3$. This evolution is dictated by a semiclassical Boltzmann equation, which incorporates the electric field and collision effects. These models are presented in Markovich *et al.* [38]; lately, there have been some analytical results regarding existence of solutions for some boundary value problems (see Poupaud [47]), but these models remain costly in terms of numerical solutions.

Finally, there is the fluid level of modeling. First is the one based on the drift diffusion equations of parabolic type for the concentration of both carriers (electrons and holes). These models can be rigorously derived from the kinetic formulation under the assumption of low electric fields (see Golse and Poupaud [31]), and have been extensively studied from the mathematical and numerical point of view (see Jerome [33] and Markowich, Ringhofer, and Schmeiser [38] and references therein). They give very good results for components whose typical length scale is of the order of a micron, but they do not seem to be valid for submicron devices or for high electric fields. Therefore, more sophisticated models have been introduced. They are based on electro-hydrodynamic equations (also called energy balanced equations or extended drift diffusion models), which are intended to take into account high field effects. They have been obtained by closing the moment equations derived from the Boltzmann equation with a phenomenological assumption on the distribution function. The distribution is assumed to be isotropic around its mean velocity (see, for instance, Blotekjaer [7], Azoff [3], Bringer and Schön [8] and Odeh, Gnudi and Rudan

[30]). Then, fluid equations are obtained with source terms modeling relaxation processes and electric fields effects coupled, with the Poisson equation. Some recent numerical work includes Chen *et al.* [12] in two space dimensions, and Gardner [26, 27], Gardner, Jerome, and Shu [28] and Fatemi *et al.* [18] in one space dimension.

Analyses addressing issues of existence, uniform bounds, and boundary layer formation for steady-state one-dimensional transonic models have been studied by Gamba [20, 22, 23]. Asher *et al.* [2] presented a phase plane analysis for some special models. The one space dimension, time-dependent, inviscid problem has been recently analyzed by Marcati and Natalini [37], Poupaud, Rascle and Vila [48], and Zhang [53]. In addition, Cordier *et al.* [14] analyzed traveling waves and jump relations for an Euler–Poisson model in the quasineutral limit. Markowich [39] treated a two-dimensional steady Euler–Poisson system in subsonic regimes, under a smallness assumption on the prescribed outflow velocity (small boundary current) and under a smallness assumption on the variation of the velocity relaxation time. Degond and Markowich [16] considered a three-dimensional potential inviscid flow model, where they proved existence and uniqueness results in a bounded domain for small Dirichlet data. Both papers deal with systems of equations that remain essentially elliptic under assumptions either on the size of the data, or on the size of the parameters under consideration.

Acknowledgements

The research of the first author is supported by the National Science Foundation under grant DMS 9623037.

Bibliography

1. Anile, A. M. and Muscato, O. (1995). An improved hydrodynamic model for carrier transport in semiconductors. *Physical Review B*, **51**, 16728–16740.
2. Ascher, U., Markowich, P., Pietra, P., and Schmeiser, C. (1991). A phase plane analysis of transonic solutions for the hydrodynamic semiconductor model. *Mathematical Models and Methods in Applied Sciences,* **1**, 347–376.
3. Azoff, E. M. (1987). Generalized energy moment equation in relaxation time approximation. *Solid State Electr.*, **30**, 913–917.
4. Baccarani, G. and Wordeman, M. R. (1985). An investigation of steady state velocity overshoot effects in Si and GaAs devices. *Solid State Electr.*, **28**, 407–416.
5. Barenger, H. U. and Wilkins, J. W. (1984). Ballistic electrons in an inhomogeneous submicron structure: thermal and contact effects. *Physical Review B*, **30**, 7349–7351.
6. Barenger, H. U. and Wilkins, J. W. (1987). Ballistic structure in the electron distribution function of small semiconducting structures: general features and specific trends. *Physical Review B,* **36**, 1487–1502.

7. Bløtekjaer, K. (1970). Transport equations for electrons in two–valley semiconductors. *IEEE Trans. Electron Devices*, **17**, 38–47.
8. Bringer, A. and Schön, G. (1988). Extended moment equations for electron transport in semiconducting submicron structures. *J. Appl. Phys.*, **61**, 2445–2554.
9. Cercignani, C., Gamba, I. M., and Levermore, C. L. (1996). A high field approximation to a Boltzmann–Poisson system in bounded domains. *Submitted for publication*.
10. Cercignani, C., Gamba, I. M., and Levermore, C. L. (1997). High field approximations to a Boltzmann–Poisson system and boundary conditions in a semiconductor. *Appl. Math. Lett.*, **10**, 111–117.
11. Chen, G.-Q. (1990). The compensated compactness method and the system of isentropic gas dynamics. *MSRI Preprint 00527-91*.
12. Chen, Z., Cockburn, B., Jerome, J. W., and Shu, C.-W. (1995). Mixed-RKDG finite element methods for the 2-D hydrodynamic model for semiconductor device simulation. *VLSI Design*, **3**, 145–158.
13. Choi, Y.-H. and Merkle, C. L. (1993). The application of preconditioning in viscous flows. *J. Comp. Phys.*, **105**, 207–223.
14. Cordier, S., Degond, P., Markowich, P. A., and Schmeiser, C. (1995). Travelling wave analysis and jump relations for Euler Poisson model in the quasineutral limit. *Asymptotic Anal.*, **11**, 209–240.
15. Courant R. and Friedrichs K. O. (1967). *Supersonic Flow and Shock Waves*. Wiley-Interscience.
16. Degond, P. and Markowich, P. A. (1993). A steady-state potential flow for semiconductors. *Ann. Mat. Pura Appl.*, **165**, 87–98.
17. Di Perna, R. J. (1985). Compensated compactness and general systems of conservation laws. *Trans. A.M.S.*, **292**, 383–420.
18. Fatemi, E., Gardner, C., Jerome, J. W., Osher, S., and Rose, D. (1991). Simulation of a steady state electron shock wave in a submicron semiconductor device using high-order upwind methods. In: *Computational Electronics* (ed. K.Hess, J. P. Leburton and U. Ravaioli), pp. 27–32. Kluwer.
19. Feistauer, M. and Nečas, J. (1985). On the solvability of transonic potential flow problem. *Z. Anal. Anwendungen*, **4**, 305–329.
20. Gamba, I. M. (1992). Stationary transonic solutions for a one-dimensional hydrodynamic model for semiconductors. *Comm. Partial Differential Equations*, **17**, 553–577.
21. Gamba, I. M. (1995). An existence and uniqueness result for a nonlinear 2-dimensional elliptic boundary value problem. *Comm. Pure Appl. Math.*, **XLVIII**, 669–689.

22. Gamba, I. M. (1992). Boundary layer formation for viscosity approximations of transonic flow. *Physics of Fluids A*, **4**, 486–490.
23. Gamba, I. M. (1994). Viscosity approximating solutions to ODE systems that admit shocks, and their limits. *Advances in Applied Mathematics*, **15**, 129–182.
24. Gamba, I. M. (1997). Sharp uniform bounds for steady potential fluid–Poisson systems. *Proceedings of the Royal Society of Edinburgh Sect. A*, **127**, 479–516.
25. Gamba, I. M. and Morawetz, C. S. (1996). A viscous approximation for a 2-D steady semiconductor or transonic gas dynamic flow: existence theorem for potential flow. *Comm. Pure Appl. Math.*, **49**, 999–1049.
26. Gardner, C. L. (1991). Numerical simulation of a steady-state electron shock wave in a submicron semiconductor device. *IEEE Trans. Electr. Devices*, **38**, 392–398.
27. Gardner, C. L. (1993). Hydrodynamic and Monte Carlo simulation of an electron shockwave in a one micrometer $n^+ - n - n^+$. *IEEE Trans. Electr. Devices*, **40**, 455–457.
28. Gardner, C. L., Jerome, J. W., and Shu, C.-W. (1993). The ENO method for the hydrodynamic model for semiconductor devices. In: *High Performance Computing: Grand Challenges in Computer Simulation* (ed. A. Tentner), pp. 96–101. The Society for Computer Simulation, San Diego.
29. Gittel, H. P. (1995). A variational approach to boundary value problems for a class of nonlinear systems of mixed type. *Z. Anal. Anwendungen*, **14**, 575–591.
30. Gnudi, A., Odeh, F., and Rudan, M. (1990). Investigation of nonlocal transport phenomena in small semiconductor devices. *European Trans. on Telecommunications and Related Technologies*, **1**, 307–312.
31. Golse, F. and Poupaud, F. (1992). Limite fluide des eáquations de Boltzmann des semi-conducteurs pour une statistique de Fermi-Dirac. *Asymptotic Analysis*, **6**, 135–160.
32. Jameson, A. (1988). Computational transonics, *Comm. Pure Appl. Math.*, **XLI**, 507–549.
33. Jerome, J. W. (1996). *Analysis of Charge Transport: a Mathematical Study of Semiconductor Devices*. Springer.
34. Klouček, P. (1994). On the existence of the entropic solution for the transonic flow problem by the method of stabilization. *Nonlinear Analysis*, **22**, 467–480.
35. Lions, P. L., Perthame, B., and Tadmor, E. (1994). Kinetic formulation of the isentropic gas dynamics and p-systems. *Comm. Math. Phys.*, **163**, 415–431.

36. Lions, P. L., Perthame, B., and Souganidis, E. (1996). Existence and stability of entropy solutions for the hyperbolic systems of isentropic gas dynamics in Eulerian and Lagrangian coordinates, *Comm. Pure Appl. Math.*, **49**, 599–638.
37. Marcati, P. and Natalini, R. (1995). Weak solutions to a hydrodynamic model for semiconductors: the Cauchy problem. *Proc. Roy. Soc. Edinburgh Sect. A* , **125**, 115–131.
38. Markowich, P., Ringhofer, C. A., and Schmeiser, C. (1989). *Semiconductor Equations.* Springer.
39. Markowich, P. (1991). On steady state Euler-Poisson models for semiconductors. *Z. Angew. Math. Phys.*, **22**, 389–407.
40. Morawetz, C. S. (1985). On a weak solution for a transonic flow problem. *Comm. Pure Appl. Math.*, **38**, 797–817.
41. Morawetz, C. S. (1995). On steady transonic flow by compensated compactness. *Methods and Applications of Analysis*, **2**, 257–268.
42. Murat, F. (1978). Compacité par compensation. *Ann. Scuola Norm. Sup. Pisa*, **5**, 489–507.
43. Nečas J. and Klouček P. (1990). The solution of transonic flow problems by the method of stabilization. *Applicable Analysis,* **37**, 143–167.
44. Nigro, N., Storti, M., Idelsohn, S. and Tezduyar, T. (1996). Physics based GMRES preconditioner for compressible and incompressible Navier–Stokes equations. *Preprint.*
45. Poupaud, F. (1992). Runaway phenomena and fluid approximation under high fields in semiconductor kinetic theory. *Z. Angew. Math. Mech.*, **72**, 359–372.
46. Poupaud, F. (1991). Derivation of a hydrodynamic systems hierarchy from the Boltzmann equation. *Appl. Math. Lett.*, **4**, 75–79.
47. Poupaud, F. (1994). Mathematical theory of kinetic equations for transport modelling in semiconductors. In: *Advances in Kinetic Theory and Computing,* Ser. Adv. Math. Appl. Sci. (ed. B. Perthame), pp. 141–168, vol.22, World Sci. Publishing, River Edge, NJ.
48. Poupaud, F., Rascle, M., and Vila, J.-P. (1995).Global solutions to the isothermal Euler-Poisson system with arbitrary large data. *Jour. of Diff. Equations.*, **123**, 93–121.
49. Serrin, J. (1959). Mathematical principles of classical fluid dynamics. In: *Handbuch der Physik,*, pp. 125–263. Springer-Verlag.
50. Stichel, P. C. and Strothmann, D. (1994). Asymptotic analysis of the high field semiconductor Boltzmann equation. *Physica A*, **202**, 553–576.
51. Synge, J. L. (1955). The motion of a viscous fluid conducting heat. *Quart. Appl. Math.*, **13**, 271–278.

52. Tartar, L. C. (1979). Compensated compactness and applications to partial differential equations. In: *Nonlinear Analysis and Mechanics,* Heriot-Watt Symposium, IV, Research Notes in Mathematics, pp. 136–192. Pitman.
53. Zhang, B. (1993). Convergence of the Godunov Scheme for a simplified one dimensional hydrodynamic model for semiconductor devices. *Commun. Math. Phys.*, **157**, 1–22.

6

Two-Carrier Semiconductor Device Models with Geometric Structure and Symmetry Properties

Gui-Qiang Chen and Joseph W. Jerome
Northwestern University

and

Chi-Wang Shu
Brown University

and

Dehua Wang
University of California at Santa Barbara

Abstract

We introduce a two-carrier (electro) hydrodynamic model, which incorporates higher-dimensional geometric effects into a one-dimensional model. A rigorous mathematical analysis is carried out for the evolution of the system in the case of piezotropic flow, including realistic carrier coupling. The proofs are constructive in nature, making use of generalized Godunov schemes with a novel fractional step, steady-state component, and compensated compactness. Two important applications are studied. We simulate: (1) the GaAs device in the notched oscillator circuit; and (2) a MESFET channel and its steady-state symmetries. The first of these applications is the well known Gunn oscillator, and we are able to replicate Monte-Carlo simulations, based upon the Boltzmann equation. For the second application, we observe the effect of a symmetry-breaking parameter, the potential bias on the drain. For those wishing to bypass the mathematical detail upon a first reading so as to concentrate upon the modeling and simulation, it is suitable to proceed directly to Sections 4 and 5 following Section 1.2.

1 Introduction

In previous work, we have studied one-and two-dimensional semiconductor devices over a wide range of parameters, via the hydrodynamic model. Important characteristics of this model include heat conduction, relaxation, and electrical forcing and heating terms. In particular, carrier transport occurs in a self-consistent electric field. The model is decidedly more complex than the standard gas dynamics model, and therefore permits more diverse solution behavior. In [18]–[20], n^+–n–n^+ diodes in one-dimension and MESFETS in two-dimensions were simulated via an essentially nonoscillatory shock capturing algorithm (ENO). In this work, we allow for the additional generalization of multi-species and geometric source terms. Such generality is driven by significant applications. For example, it has become customary among device physicists to differentiate the same particle carrier (say, electrons) on the basis of its energy valley occupancy. Monte-Carlo simulations of the Boltzmann equation routinely proceed in this manner. Thus, we are able to simulate two-valley gallium arsenide diodes, and examine their oscillatory behavior when coupled to simple circuits. This application is called the Gunn oscillator in the device literature. Also, we are able to examine the symmetry-breaking properties of the potential bias on the drain of a MESFET transistor, and compare in detail the solutions of the two-dimensional model with those of the one-dimensional model with spherical symmetry.

We present a comprehensive mathematical analysis for the reduced version of the multi-species model with geometric source terms, viz., the piezotropic model, with pressure a specific function of density. In the remainder of the introduction, we briefly describe the two applications, and summarize our mathematical results. A final comment is in order for the reader of this paper. Those not interested in the mathematical analysis can pass on to Sections 4 annd 5, after familiarizing themselves with the applications of Section 1.1 and Section 1.2. On the other hand, the reader interested in the mathematical proofs contained in Section 2 and in Section 3 can pass now to Section 1.3. We have presented these proofs, which are not available anywhere else.

1.1 Description of the Gunn Oscillator

The equations describing an RLC tank circuit, connected to a Gunn diode, are:

$$V_D(t) = V_B - L\frac{dI(t)}{dt}, \quad I(t) = I_d(t) + C\frac{dV_D(t)}{dt} + \frac{V_D(t)}{R}, \tag{1.1}$$

where $V_D(t)$ is the voltage at the device terminal, V_B is the bias voltage, $I(t)$ is the current flowing through the battery, and C is the total capacitance, which includes the so-called cold capacitance. $I_d(t)$ is the particle current, which is spatially constant throughout the diode. In [22], a Monte-Carlo simulation of the Boltzmann equation was used to update $I_d(t)$. Earlier, a single valley hydrodynamic model was used in simulation by the authors of [9], whereas here we employ a two-valley hydrodynamic model. The coupling terms and the system have the structure of [1]. As derived in [17], each carrier in the diode then satisfies a system of the form:

$$\begin{cases} \partial_t \rho + \nabla \cdot (\rho v) = C_\rho, \\ \partial_t m + v(\nabla \cdot m) + (m \cdot \nabla)v = -\frac{e}{m^*}\rho F - \nabla(\rho k_b T/m^*) + C_m, \\ \partial_t E + \nabla \cdot (v\, E) = -\frac{e}{m^*}\rho v \cdot F - \nabla \cdot (v\rho k_b T/m^*) + \nabla \cdot (\kappa \nabla T) + C_E. \end{cases} \tag{1.2}$$

Here, ρ denotes particle mass density, related to concentration n and effective mass m^* via $\rho = m^*n$, m denotes particle momentum density, related to velocity v through $m = \rho v$, and E the mechanical energy density. F denotes the electric field, $\mathcal{T}$ the carrier temperature, e the charge modulus, κ the heat conductivity, k_b Boltzmann's constant, and C_ρ, C_m, and C_E denote relaxation expressions. The systems are coupled through the Poisson electrostatic equation as well. We shall give greater detail in Section 4. There, we shall interpret the two copies of (1.2) as describing GaAs electron carriers associated with lower (Γ) and middle (L) energy valleys. The occasional use of the third (upper) valley is not made here.

1.2 Basic MESFET Description

Next we describe a two-dimensional MESFET of the size 0.6×0.2 μm^2. The source and the drain each occupies 0.1 μm at the upper left and the upper right, respectively, with a gate occupying 0.2 μm at the upper middle (Fig. 1). The doping is defined by $n_d = 3 \times 10^5$ μm^{-3} in $[0, 0.1] \times [0.15, 0.2]$ and in $[0.5, 0.6] \times [0.15, 0.2]$, and $n_d = 1 \times 10^5$ μm^{-3} elsewhere. We apply, at the drain, voltage biases varying up to $vbias = 2$ V. This bias has been described earlier as a symmetry-breaking parameter, and we shall investigate this in detail in the sequel. The gate is a Schottky contact, with negative voltage bias up to $vgate = -0.8$ V and very low concentration value $n = 3.8503 \times 10^{-8}$ μm^{-3} (following Selberherr [27]). The lattice temperature is taken as $\mathcal{T}_0 = 300$ K. The mathematical model for the MESFET is the *steady-state version* of system (1.2), coupled to Poisson's electrostatic equation.

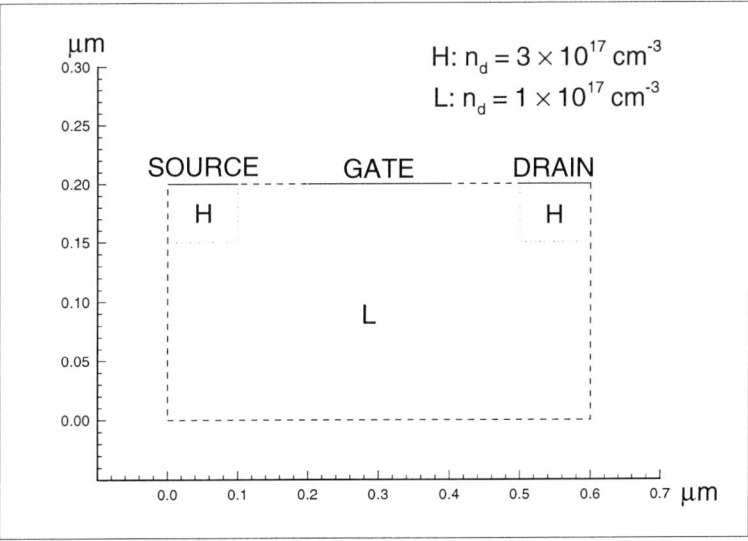

FIG. 1. Two-dimensional MESFET. The geometry and the doping n_d.

1.3 A Well-Posed Reduced Model

Consider a reduced model, the compressible, two-carrier, Euler–Poisson equations:

$$\begin{cases} \partial_t \rho_i + \nabla \cdot \vec{m}_i = R_i(\rho_1, \rho_2), \\ \partial_t \vec{m}_i + \nabla \cdot \left(\frac{\vec{m}_i \otimes \vec{m}_i}{\rho_i} \right) + \nabla p(\rho_i) = \rho_i \nabla \phi - \frac{\vec{m}_i}{\tau_i} + \vec{H}_i(\rho_1, \rho_2, E_1, E_2), \quad (1.3) \\ \Delta \phi = \rho_1 + \rho_2 - n_d(\vec{x}), \quad i = 1, 2, \quad \vec{x} \in \mathbb{R}^N, \end{cases}$$

where $\rho_i(\vec{x}, t)$, $\vec{m}_i(\vec{x}, t)$, and $\phi(\vec{x}, t)$ denote the density, the momentum, and the potential of the flows, respectively, and $p(\rho_i) = \rho_i^\gamma / \gamma$, $\gamma > 1$, is the pressure, $E_i = \frac{\rho_i^{\gamma-1}}{\gamma(\gamma-1)} + \frac{|\vec{m}_i|^2}{2\rho_i^2}$ is the mechanical energy, $\tau_i > 0$ is the momentum relaxation time, and $n_d(\vec{x})$ is the doping profile. For simplicity, in this mathematical model, we have selected units in which e/m_i^* has been absorbed into the units of ρ_i, $i = 1, 2$, the dielectric constant has been absorbed into the units of ϕ, and the charge modulus has been absorbed into the units of n_d. The initial-boundary problem for the system (1.3) with geometric symmetry is ($1 < x < 2$, $t > 0$):

$$\begin{cases} \partial_t \rho_i + \partial_x m_i = a(x) m_i + R_i(\rho_1, \rho_2), \\ \partial_t m_i + \partial_x \left(\frac{m_i^2}{\rho_i} + p(\rho_i) \right) = a(x) \frac{m_i^2}{\rho_i} + \rho_i \phi_x - \frac{m_i}{\tau_i} + H_i(\rho_1, \rho_2, E_1, E_2), \\ \phi_{xx} = a(x) \phi_x + \rho_1 + \rho_2 - n_d(x), \quad i = 1, 2, \end{cases} \quad (1.4)$$

$$\begin{cases} (\rho_i, m_i)|_{t=0} = (\rho_{i0}(x), m_{i0}(x)), \\ m_i|_{x=1} = m_i|_{x=2} = 0, \quad \phi|_{x=1} = \phi_1(t) \in L^\infty, \quad \phi|_{x=2} = \phi_2(t) \in L^\infty, \end{cases} \quad (1.5)$$

where the field term ϕ_x is nonlocal (self-consistent) and $a(x)$ is a C^1 function that can be represented by $a(x) = -A'(x)/A(x)$. The function $A(x)$ describes the cross-sectional area at x in a variable-area duct such as a nozzle channel, and $A(x) = \frac{2\pi^{N/2}}{\Gamma(N/2)} x^{N-1}$ for spherically symmetric flow in N dimensions, such as in the MESFET, for the one-carrier case we test.

The Euler–Poisson equations for two carriers with $a(x) = 0$ have been studied for some special couplings: The case $R_i = H_i = 0$ in [26] by the Godunov scheme with fractional step techniques and the case $R_i = (1 - \rho_1 \rho_2) Q(\rho_1, \rho_2)$, $H_i = 0$, $0 \leq Q(\rho_1, \rho_2) \leq \frac{Q_0}{1+\rho_1+\rho_2}$ in [14] by the viscosity method. The system for one carrier with general $a(x) \in C^1$ is solved in [8].

We develop a new shock capturing numerical scheme and apply this scheme to construct global entropy solutions to the system (1.4)–(1.5) with nonzero $a(x)$ and general R_i and H_i. More precisely, we consider the following coupling terms $R_i(\rho_1, \rho_2)$ and $H_i(\rho_1, \rho_2, E_1, E_2)$:

(A1) R_i and H_i are Lipschitz functions in the variables $\rho_1 \geq 0$, $\rho_2 \geq 0$, E_1, and E_2.

(A2) There exist a constant $C > 0$ and a decomposition of R_i: $R_i(\rho_1, \rho_2) = R_i^+(\rho_1, \rho_2) - R_i^-(\rho_1, \rho_2)$ with $R_i^\pm(\rho_1, \rho_2) \geq 0$ such that, for all $\rho_1, \rho_2 > 0$

and $i = 1, 2$,

$$R_i(\rho_1, \rho_2) \leq C, \qquad |H_i(\rho_1, \rho_2, E_1, E_2)| \leq C\rho_i;$$
$$R_i^+(\rho_1, \rho_2) \leq C\rho_i, \qquad \text{if } R_i^+(\rho_1, \rho_2) \geq R_i^-(\rho_1, \rho_2) \text{ and } \rho_i \geq c_0;$$
$$0 < R_i^-(\rho_1, \rho_2) \leq C\rho_i, \text{ if } R_i^+(\rho_1, \rho_2) < R_i^-(\rho_1, \rho_2),$$

where $c_0 = (\theta/(\theta + 1))^{1/\theta}$ with $\theta = (\gamma - 1)/2$.

We then have the following theorem, which is a synopsis of Theorems 3.9 and 3.10 of the sequel:

Theorem 1.1 *Let $a(x)$ be a C^1 function and $1 < \gamma \leq 5/3$. Let $R_i(\rho_1, \rho_2)$ and $H_i(\rho_1, \rho_2, E_1, E_2)$ satisfy assumptions (A1)-(A2). Then there exists a sequence of approximate solutions $(\rho_i^h(x, t), m_i^h(x, t))$, for $i = 1, 2$, converging a.e. to an entropy solution $(\rho_i(x, t), m_i(x, t))$, of (1.4)-(1.5) such that $0 \leq \rho_i(x, t) \leq C(T) < \infty$, $|m_i(x, t)/\rho_i(x, t)| \leq C(T) < \infty$, for $0 \leq t \leq T < \infty$, $x \in (1, 2)$, a.e.*

Assumptions (A1)–(A2) are in fact quite general. For example, $R_i = 0$, $R_i = \frac{1 - \rho_1 \rho_2}{1 + \rho_1 + \rho_2}$, $R_i = \frac{(-1)^i (\rho_1 - \rho_2)}{1 + \rho_1 + \rho_2}$, $H_i = 0$, and $H_i = \frac{\rho_i E_i}{1 + E_1 + E_2}$ are in this class.

Since $a(x)$ is not equal to zero, the nonlinear resonance between characteristic modes and the geometric source terms occurs at the sonic state, that is, some of the characteristic speeds and the source speed coincide at the sonic state. Such a nonlinear resonance causes extra difficulties (cf. [5, 6, 13, 15, 23, 24, 16]). Recently, an efficient shock capturing scheme was developed in [5, 6] to solve the Euler equations with geometric structure by incorporating the steady-state solutions with the Godunov scheme. Our problem (1.4) and (1.5) involves both the geometric source terms and nonlocal source terms.

Due to the geometric source terms, we adopt the approach of Chen and Glimm [5, 6]. One of the key ideas of this approach is to use the piecewise approximate steady-state solutions, which incorporate the geometric source terms, to replace the piecewise constants from the Riemann solutions as the building blocks. The main difficulty in achieving this is that, in the transonic case, no smooth steady-state solution exists, and an approximate steady-state solution, including a standing shock, has to be introduced, satisfying some important properties similar to those of the smooth solution in each cell: (a) the oscillation of the steady-state solution around the Godunov value must be of the same order as the cell length to obtain the L^∞ estimate for the convergence arguments; (b) the difference between the average of the steady-state solution over each cell and the Godunov value must be higher than first-order in the cell length to ensure the consistency between the corresponding approximate solutions and the Euler equations. These requirements are naturally satisfied by smooth steady-state solutions that are bounded away from the sonic state in the cell. The sonic difficulty is overcome, as in computational physics, by using the additional standing shock with continuous mass and adjusting its left state and right state in the density and its location to control the growth of the density. This construction considerably improves the traditional Godunov scheme for this case.

Due to the nonlocal source terms, we also incorporate the fractional step procedure into our construction of approximate solutions with the steady-state solutions as their fundamental building blocks. First we solve Poisson's equation to get ϕ_x, which is the nonlocal term. To obtain the uniform bound of the approximate solutions, we estimate the Riemann invariants, for which the nonlocal term is involved. For the case $a(x) = 0$, one has the conservation of particles. For the case $a(x) \neq 0$, one does not have such a conservation principle because of the geometric source terms. Therefore we have to make a proper estimate on the nonlocal term in order to get the uniform estimate of the approximate solutions. For this purpose, we use the conservation of mass in a different but equivalent way. We change the definition of the Godunov values in the scheme and prove a new property of the steady-state solution. Then we can make a new estimate on the nonlocal term, which is sufficiently robust so that the uniform estimate can be achieved. To estimate the H^{-1} compactness of the weak entropy dissipation measures, we first make the estimates on the mechanical entropy pair, which will also be used later to prove the convergence and existence. Because of the different definition of the Godunov values, we must prove the existence separately for the first and the second equations of (1.4), and use the new property of the steady-state solution. Some extra terms from the fractional step procedure must be taken into account.

These requirements enable us to deduce the strong convergence of the approximate solutions with the aid of a compactness framework (see Chen [2, 3]). The framework takes the vacuum into account in correct physical variables (ρ, m) near the vacuum, rather than the variables (ρ, u) ($u :=$ velocity) that are physically incorrect on the vacuum. The compactness framework we use was proved in [12] for the case $\gamma = 1 + \frac{2}{2\ell+1}$, $\ell \geq 2$, and in [2, 11] (also see [3]) for the general case of gases with $1 < \gamma \leq 5/3$. Finally, the new existence theorem for the global weak solution to the initial-boundary problems of (1.3)–(1.4), with nonlocal source terms, is established with the aid of this framework.

2 The Mathematical Framework

Here we provide the framework for the proof of Theorem 1.1. The proof is given in Section 3.

2.1 Preliminaries

In this section, we review some basic facts about the Riemann solutions for homogeneous systems and the steady-state solutions.

Consider the homogeneous system:

$$u_t + f(u)_x = 0, \quad 1 < x < 2, \tag{2.1}$$

where $u = (\rho, m)^\top$ and $f(u) = (m, \frac{m^2}{\rho} + p(\rho))^\top$ with $p(\rho) = \rho^\gamma/\gamma$, $\gamma > 1$.

The eigenvalues are
$$\lambda_1 = \frac{m}{\rho} - \rho^\theta, \quad \lambda_2 = \frac{m}{\rho} + \rho^\theta,$$

where $\theta = \frac{\gamma-1}{2}$. The two characteristic fields are genuinely nonlinear.
The Riemann invariants are
$$w = \frac{m}{\rho} + \frac{\rho^\theta}{\theta}, \quad z = \frac{m}{\rho} - \frac{\rho^\theta}{\theta}.$$

The discontinuity in the weak solution of (2.1) satisfies the Rankine–Hugoniot condition:
$$\sigma(u - u_0) = f(u) - f(u_0), \tag{2.2}$$

where σ is the propagation speed of the discontinuity, and u_0 and u are the corresponding left state and right state, respectively. A discontinuity is a shock if it satisfies the entropy condition:
$$\sigma(\eta(u) - \eta(u_0)) - (q(u) - q(u_0)) \geq 0, \tag{2.3}$$

for any convex entropy pair (η, q). The shock with speed $\sigma = 0$ is called the standing shock.

Consider the Riemann problem consisting of (2.1) with initial data,
$$u|_{t=0} = \begin{cases} u_-, & x < x_0, \\ u_+, & x > x_0, \end{cases} \tag{2.4}$$

where $x_0 \in (1, 2)$, $u_\pm = (\rho_\pm, m_\pm)^\top$, and $\rho_\pm \geq 0$ and $m_\pm$ are constants satisfying $\left|\frac{m_\pm}{\rho_\pm}\right| < \infty$. There are two distinct types of rarefaction waves and shock waves.

For the Riemann problem with data (2.4) and the Riemann initial-boundary problem of (2.1) with data:
$$u|_{t=0} = u_+, \quad m|_{x=1} = 0, \tag{2.5}$$

we have the following facts regarding the solutions.

Lemma 2.1 *There exists a piecewise smooth entropy solution $u(x, t)$ for each of the problems (2.4) and (2.5), respectively, satisfying*
$$\begin{cases} w(u(x,t)) \leq \max(w(u_-), w(u_+)), \\ w(u(x,t)) - z(u(x,t)) \geq 0, \end{cases} \tag{2.6}$$

and, for (2.4),
$$z(u(x,t)) \geq \min(z(u_-), z(u_+)),$$

and, for (2.5),
$$z(u(x,t)) \geq \min(z(u_+), 0).$$

Lemma 2.2 *If w_0, z_0 are the bounds implied in Lemma 2.1, then for the Riemann problem (2.4), the region*

$$\Sigma = \{(\rho, m) : w \leq w_0, z \geq z_0, w - z \geq 0\}$$

is an invariant region of (2.1). For the Riemann initial-boundary problem (2.5), the region

$$\Sigma = \{(\rho, m) : w \leq w_0, z \geq z_0, w - z \geq 0\}, \quad z_0 \leq 0 \leq \frac{w_0 + z_0}{2},$$

is an invariant region of (2.1). That is, if the Riemann data lie in Σ, then the Riemann solutions $u(x,t) \in \Sigma$ and $\frac{1}{b-a}\int_a^b u(x,t)dx \in \Sigma$.

For the Riemann initial-boundary problem of (2.1) with data:

$$u|_{t=0} = u_-, \quad m|_{x=2} = 0, \tag{2.7}$$

we have the similar results to those for (2.5) in the above two lemmas.

A pair of mappings $(\eta, q) : \mathbb{R}^2 \to \mathbb{R}^2$ is called an entropy–entropy flux pair if $\nabla q = \nabla \eta \nabla f$. If $\tilde{\eta}(\rho, v) \equiv \eta(\rho, \rho v)$ satisfies $\tilde{\eta}(0, v) = 0$, for any fixed $v = \frac{m}{\rho}$, then η is called a weak entropy. For example, the mechanical energy-energy flux pair

$$\eta_* = \frac{1}{2}\frac{m^2}{\rho} + \frac{1}{\gamma(\gamma-1)}\rho^\gamma, \quad q_* = m\left(\frac{1}{2}\frac{m^2}{\rho^2} + \frac{\rho^{\gamma-1}}{\gamma-1}\right), \tag{2.8}$$

is a strictly convex weak entropy pair for (2.1).

Lemma 2.3 *Assume $(\rho, m)^\top = (\rho, \rho v)^\top$ is a Riemann solution of (2.1) satisfying $0 \leq \rho \leq C'$, $|v| \leq C'$, for some constant $C' > 0$; then there exists a constant $C > 0$ such that*

$$|\nabla \eta| \leq C, \quad |\nabla q| \leq C; \quad |u^\top \nabla^2 \eta u| \leq C u^\top \nabla^2 \eta_* u; \quad |\sigma[\eta] - [q]| \leq C(\sigma[\eta_*] - [q_*]),$$

for any weak entropy pair (η, q), where u is any vector and the constant C is independent of u. Here, $\nabla^2 \eta$ denotes the Hessian matrix.

Next we revisit some important properties of the steady-state solutions (see [5, 6, 8]). Consider the system of steady-state equations with boundary condition:

$$\begin{cases} f(u)_x = a(x)g(u), \\ u|_{x=x_0} = u_0, \end{cases} \tag{2.9}$$

where

$$u = (\rho, m)^\top, \quad u_0 = (\rho_0, m_0)^\top,$$

$$f(u) = (m, \frac{m^2}{\rho} + p(\rho))^\top, \quad g(u) = (m, \frac{m^2}{\rho})^\top,$$

and
$$a(x) = -\frac{A'(x)}{A(x)}, \quad \text{with} \quad A(x) \in C^2, A(x) \geq c_0 > 0.$$

Set the sound speed: $c = \rho^\theta$. Then $M = M(u(x)) = \frac{v(x)}{c(x)}$ is the Mach number, and $M_0 = M(u_0)$.

For the nonsonic case, $|M_0^2 - 1| \geq h^\beta M_0^2$, with $\beta \in (0, \frac{1}{6})$, $h \in (0, h_0)$ for some sufficiently small $h_0 \in (0, 1)$, (2.9) has a smooth solution.

When $|M_0^2 - 1| < h^\beta M_0^2$, the steady-state equation (2.9) does not have exact smooth solutions, but has approximate solutions satisfying

$$|f(u)_x - a(x)g(u)| \leq o(1), \quad \text{as} \quad h \to 0. \tag{2.10}$$

Near the sonic case, $K_0\sqrt{h} \leq |M_0^2 - 1| \leq h^\beta M_0^2$, with $K_0 = 2\sqrt{\frac{\|a\|_C}{\theta}}$, take

$$\begin{cases} \rho(x) = \rho_0 \left(1 + \frac{M_0^2 - 1}{2(\theta+1)} \left(1 - \sqrt{1 - \frac{4(\theta+1)a_0}{(M_0^2-1)^2}(x - \tilde{x})}\right)\right), \\ m(x) = m_0(1 + a_0(x - x_0)), \end{cases} \tag{2.11}$$

where $a_0 = a(x_0)$, $\tilde{x} \in (x_0 - \frac{h}{2}, x_0 + \frac{h}{2})$. Then $u = (\rho, m)^\top$ is an approximate solution in the sense of (2.10) and satisfies locally,

$$f(u)_x - a(x)g(u) = O(h^\beta). \tag{2.12}$$

For the transonic case, $|M_0^2 - 1| < K_0\sqrt{h}$, we introduce a standing shock at $x = \tilde{x}$ with left state $u_- = (\rho_{0-}, m_0)$ and right state $u_+ = (\rho_{0+}, m_0)$, where

$$\rho_{0\pm} = \rho_0 \left(\frac{\theta M_0 + 1}{\theta + 1}\right)^{\frac{1}{\theta}} (1 \pm K_0\sqrt{h}).$$

The corresponding Mach numbers are $M_{0\pm}^2 = 1 \mp 2(\theta + 1)K_0\sqrt{h} + O(h)$. Take

$$\rho_\pm(x) = \rho_{0\pm} \left(1 + \frac{M_{0\pm}^2 - 1}{2(\theta+1)} \left(1 - \sqrt{1 - \frac{4(\theta+1)a_0}{(M_{0\pm}^2-1)^2}(x - \tilde{x})}\right)\right), \tag{2.13}$$

with $\tilde{x} \in (x_0 - \frac{2+\theta}{4(1+\theta)}h, x_0 + \frac{2+\theta}{4(1+\theta)}h)$. Then $u(x) = (\rho(x), m(x))^\top$ defined by

$$\begin{cases} \rho(x) = \begin{cases} \rho_-(x), & x \in [x_0 - \frac{h}{2}, \tilde{x}), \\ \text{standing shock}, & x = \tilde{x}, \\ \rho_+(x), & x \in (\tilde{x}, x_0 + \frac{h}{2}], \end{cases} \\ m(x) = m_0(1 + a_0(x - x_0)), \end{cases} \tag{2.14}$$

is an approximate solution of (2.9) with $\rho_0 \geq 0$ in the sense of (2.10) satisfying (2.12). Furthermore, we have

Lemma 2.4 There exists a smooth steady-state solution $u(x)$ of (2.9) when $|M_0^2 - 1| \geq h^\beta M_0^2$, an approximate smooth steady-state solution $u(x)$ when $K_0\sqrt{h} \leq |M_0^2 - 1| \leq h^\beta M_0^2$, and an approximate steady-state solution including a standing shock at some $\tilde{x} \in (x_0 - \frac{2+\theta}{4(1+\theta)}h, x_0 + \frac{2+\theta}{4(1+\theta)}h)$ when $|M_0^2 - 1| < K_0\sqrt{h}$, with $h \leq h_0$, in the sense of (2.10) such that, for $x \in [x_0 - \frac{h}{2}, x_0 + \frac{h}{2}]$,

$$\begin{cases} \rho(x) \geq 0, \\ u(x) = u_0(1 + O(\sqrt{h})), \end{cases} \tag{2.15}$$

$$\begin{cases} w(u(x)) \leq w(u_0)(1 + Ch), & \text{if } M_0 > 0, \\ z(u(x)) \geq z(u_0)(1 + Ch), & \text{if } M_0 < 0, \end{cases} \tag{2.16}$$

$$\frac{1}{h}\int_{x_0-\frac{h}{2}}^{x_0+\frac{h}{2}} u(x)dx = u_0(1 + O(h^{2(1-\beta)})), \tag{2.17}$$

and

$$\frac{1}{h}\int_{x_0-\frac{h}{2}}^{x_0+\frac{h}{2}} A(x)(\rho(x) - \rho_0)dx = \rho_0 O(h^{1+\beta}), \tag{2.18}$$

where the constant C and the bounds $O(\sqrt{h})$, $O(h^{2(1-\beta)})$, and $O(h^{1+\beta})$ depend only on the bound of $A(x)$ and are independent of M_0, and $h_0 > 0$ is sufficiently small.

2.2 The Shock Capturing Scheme

Consider the following problem:

$$\begin{cases} u_t + f(u)_x = a(x)g(u) + G(u, x, t), & 1 < x < 2, \\ u|_{t=0} = u_0(x), \\ m|_{x=1} = m|_{x=2} = 0, \end{cases} \tag{2.19}$$

where $u = (\rho, m)^\top$, $f(u)$, $g(u)$, and $a(x)$ are the same as in (2.9), and $G = (G_1, G_2) \in C$.

In this section, we construct the approximate solutions $u^h = (\rho^h, m^h)^\top = (\rho^h, \rho^h v^h)^\top$ of (2.19) in the strip $0 \leq t \leq T$ for any fixed $T \in (0, \infty)$, where $h = \frac{1}{M} > 0$, M a large positive integer, and $\Delta t > 0$ are the space mesh length and the time mesh length, respectively, and satisfy the following Courant–Friedrichs–Levy condition:

$$\Lambda = \max(\sup_{0 \leq t \leq T} |\lambda_k(\rho^h, m^h)|) \leq \frac{\gamma - 1}{4(\gamma + 1)}\frac{h}{\Delta t} \leq 2\Lambda,$$

where λ_k, $k = 1, 2$, are the eigenvalues of (2.19).

Assume that $u^h(x,t)$ is defined for $t < n\Delta t$. Then we define $u_j^n = (\rho_j^n, m_j^n)$ as:

$$\rho_j^n = \frac{\int_{1+(j-\frac{1}{2})h}^{1+(j+\frac{1}{2})h} A(x)\rho^h(x, n\Delta t - 0)dx}{\int_{1+(j-\frac{1}{2})h}^{1+(j+\frac{1}{2})h} A(x)dx}, \quad m_j^n = \frac{1}{h}\int_{1+(j-\frac{1}{2})h}^{1+(j+\frac{1}{2})h} m^h(x, n\Delta t - 0)dx,$$

for $2 \leq j \leq M - 2$; and

$$\rho_1^n = \frac{\int_1^{1+\frac{3}{2}h} A(x)\rho^h(x, n\Delta t - 0)dx}{\int_1^{1+\frac{3}{2}h} A(x)dx}, \quad m_1^n = \frac{2}{3h}\int_1^{1+\frac{3}{2}h} m^h(x, n\Delta t - 0)dx;$$

$$\rho_{M-1}^n = \frac{\int_{2-\frac{3}{2}h}^{2} A(x)\rho^h(x, n\Delta t - 0)dx}{\int_{2-\frac{3}{2}h}^{2} A(x)dx}, \quad m_{M-1}^n = \frac{2}{3h}\int_{2-\frac{3}{2}h}^{2} m^h(x, n\Delta t - 0)dx.$$

In the strip $n\Delta t \leq t < (n+1)\Delta t$, we shall construct $u_0^h(x,t)$ as follows:
(a) For $1 + jh \leq x \leq 1 + (j+1)h, 1 \leq j \leq M - 2$, $u_0^h(x,t)$ is an approximate solution, as described below, of the generalized Riemann problem for the system,

$$u_t + f(u)_x = a(x)g(u), \tag{2.20}$$

with initial data,

$$u|_{t=n\Delta t} = \begin{cases} u_-(x), & x < 1 + (j+\frac{1}{2})h, \\ u_+(x), & x > 1 + (j+\frac{1}{2})h, \end{cases}$$

where $u_-(x)$ and $u_+(x)$ are smooth solutions or approximate solutions of the steady-state equation,

$$f(u)_x = a(x)g(u), \tag{2.21}$$

with boundary conditions: $u_-(1 + jh) = u_j^n$, $u_+(1 + (j+1)h) = u_{j+1}^n$ in the sense of Section 2.1.

(b) For $1 \leq x \leq 1 + h$, $u_0^h(x,t)$ is an approximate solution of the generalized Riemann initial-boundary problem of (2.20) with data:

$$u|_{t=n\Delta t} = u_1^+(x), \quad m|_{x=1} = 0, \tag{2.22}$$

where $u_1^+(x)$ is the smooth solution or the approximate solution of the steady-state equation (2.21) with boundary condition: $u_1^+(1 + h) = u_1^n$ in the sense of Section 2.1.

(c) For $2 - h \leq x \leq 2$, $u_0^h(x,t)$ is an approximate solution of the generalized Riemann initial-boundary problem of (2.20) with data:

$$u|_{t=n\Delta t} = u_{M-1}^-(x), \quad m|_{x=2} = 0, \tag{2.23}$$

where $u_{M-1}^-(x)$ is the smooth solution or the approximate solution of the steady-state equation (2.21) with boundary condition: $u_{M-1}^-(2 - h) = u_{M-1}^n$ in the sense of Section 2.1.

The approximate solution of these problems is now described. We solve the above problem for small time *approximately* to obtain $u_0^h(x,t)$ by perturbing about the solution R of the corresponding Riemann problem of the homogeneous system:
$$u_t + f(u)_x = 0, \qquad (2.24)$$
with data
$$u|_{t=n\Delta t} = \begin{cases} u_-(1+(j+\tfrac{1}{2})h-0), & x < 1+(j+\tfrac{1}{2})h, \\ u_+(1+(j+\tfrac{1}{2})h+0), & x > 1+(j+\tfrac{1}{2})h, \end{cases}$$

for $1+jh \leq x < 1+(j+1)h$, $1 \leq j \leq M-2$; and the Riemann initial-boundary problems of (2.24) with data (2.22), for $1 < x \leq 1+h$; and with data (2.23), for $2-h \leq x < 2$.

First, let
$$R_a = \begin{cases} R = (\rho, m), & \text{if } \rho(x,t) \geq 2h^\beta, \\ (2h^\beta, m), & \text{otherwise.} \end{cases}$$

Then $R_a(x,t)$ satisfies the entropy condition on its discontinuities and
$$|R_a(x,t) - R(x,t)| \begin{cases} = 0, & \text{if } \rho(x,t) \geq 2h^\beta, \\ \leq Ch^\beta, & \text{otherwise.} \end{cases}$$

As in [10], we approximate the possible existing k-th rarefaction waves (u_-^r, u_+^r), $k = 1, 2$, in $R_a(x,t)$ by finite discontinuous rays $\frac{x_l}{t} = \lambda_k(u_l^r)$ separating finite constant states $u_l^r, l = 0, 1, \cdots, L_r$, with $u_0^r = u_-^r$ and $u_{L_r}^r = u_+^r$ such that
if $k = 1$, $w(u_{l+1}^r) = w(u_l^r) + h$, $z(u_{l+1}^r) = z(u_l^r)$, $0 \leq l \leq L_r - 1$,
if $k = 2$, $z(u_{l+1}^r) = z(u_l^r) + h$, $w(u_{l+1}^r) = w(u_l^r)$, $0 \leq l \leq L_r - 1$.
In this way, we obtain the approximate Riemann solutions consisting of finite discontinuities separating finite constant states $u_l, l = 0, 1, \cdots, L$, with $u_0 = u_-(1+(j+\tfrac{1}{2})h-0)$ and $u_L = u_+(1+(j+\tfrac{1}{2})h+0)$. Let $\hat{u}_l(x) = (\hat{\rho}_l(x), \hat{m}_l(x))$ be the exact smooth or approximate steady-state solutions such that
$$\hat{u}_l(1+(j+\tfrac{1}{2})h) = u_l.$$

We use the cut-off technique and denote by $u_l(x) = (\rho_l(x), \rho_l(x)v_l(x))$, $0 \leq l \leq L$, the approximate steady-state solutions as follows:
$$\rho_l(x) = \max(\hat{\rho}_l(x), 2h^\beta), \qquad v_l(x) = \frac{\hat{m}_l(x)}{\hat{\rho}_l(x)}, \qquad 0 \leq l \leq L.$$

The approximate solution $u_0^h(x,t) = (\rho_0^h(x,t), m_0^h(x,t))$ in the rectangle $[1+jh, 1+(j+1)h] \times [n\Delta t, (n+1)\Delta t)$ or $[1, 1+h] \times [n\Delta t, (n+1)\Delta t)$ or $[2-h, 2] \times [n\Delta t, (n+1)\Delta t)$ consists of the exact or approximate steady states $u_l(x)$, $l = 0, 1, \cdots, L$, separated by the discontinuities, subject to the Rankine–Hugoniot condition, with speeds
$$\frac{dx(t)}{dt} = u_l(x(t)) + (-1)^k \sqrt{\frac{\rho_{l+1}(x(t))}{\rho_l(x(t))} \frac{p(\rho_{l+1}(x(t))) - p(\rho_l(x(t)))}{\rho_{l+1}(x(t)) - \rho_l(x(t))}},$$

with $k = 1$ or $k = 2$ determined by the k-th original elementary waves from which the discontinuity originates. Then the approximate solutions $u_0^h(x,t)$ approach the approximate Riemann solutions as $n\Delta t \to t$.

We have the following estimates on the entropy as in [5].

Lemma 2.5 *There is a constant C depending only on the uniform bound of $u_0^h(x,t)$ such that, on any approximate shock wave with speed σ_l,*

$$\sigma_l(\eta_*(u_{l+1}) - \eta_*(u_l)) - (q_*(u_{l+1}) - q_*(u_l)) > 0,$$

and

$$|\sigma_l(\eta(u_{l+1}(x(t))) - \eta(u_l(x(t)))) - (q(u_{l+1}(x(t))) - q(u_l(x(t))))$$
$$-(\sigma_l(\eta_*(u_{l+1}) - \eta_*(u_l)) - (q_*(u_{l+1}) - q_*(u_l)))| \le Ch^{\frac{3}{2} - 2\beta};$$

and on the discontinuous rays, $x = x_l(t), \sigma_l = \frac{dx_l(t)}{dt}$, of the approximate rarefaction waves,

$$|\sigma_l(\eta(u_{l+1}(x(t))) - \eta(u_l(x(t)))) - (q(u_{l+1}(x(t))) - q(u_l(x(t))))| \le Ch^{\frac{3}{2} - 2\beta},$$

for any C^2 weak entropy-entropy flux pair (η, q) and the mechanical energy-energy flux (η_*, q_*).

Finally, we define the approximate solution $u^h(x,t) = (\rho^h(x,t), m^h(x,t))$ of (2.19) in the strip $n\Delta t \le t < (n+1)\Delta t$ by the fractional step procedure:

$$u^h(x,t) = u_0^h(x,t) + G(u_0^h(x,t), x, t)(t - n\Delta t). \tag{2.25}$$

3 Spherically Symmetric Solutions and Nozzle Solutions

Consider the spherically symmetric solutions of (1.3) in $\mathbf{R}^N$:

$$(\rho_i(\vec{x},t), \vec{m}_i(\vec{x},t), \phi(\vec{x},t)) = (\rho_i(x,t), m_i(x,t)\frac{\vec{x}}{x}, \phi(x,t)),$$

where $x = |\vec{x}|, m_i(x,t) = \rho_i(x,t)v_i(x,t)$. Then (1.3–1.4) becomes

$$\begin{cases} \partial_t u_i + \partial_x f(u_i) = a(x)g(u_i) + G(u_1, u_2, x, t), & 1 < x < 2, \ t > 0, \\ u_i|_{t=0} = u_{i0}(x), \\ m_i|_{x=1} = m_i|_{x=2} = 0, & i = 1, 2, \end{cases} \tag{3.1}$$

where

$$u_i = (\rho_i, m_i)^\top, \quad f(u_i) = (m_i, \frac{m_i^2}{\rho_i} + p(\rho_i))^\top, \quad g(u_i) = (m_i, \frac{m_i^2}{\rho_i})^\top, \quad G = (G_1, G_2)^\top,$$

and $a(x) = -\frac{N-1}{x} = -\frac{A'(x)}{A(x)}$, $A(x) = N\omega_N x^{N-1}$, $\omega_N = \frac{2\pi^{N/2}}{N\Gamma(N/2)}$, with

$$\begin{cases} G_1 = R_i(\rho_1, \rho_2), \\ G_2 = \rho_i \phi_x - \frac{m_i}{\tau_i} + H_i(\rho_1, \rho_2, E_1, E_2), \end{cases} \tag{3.2}$$

where

$$\phi_x = x^{1-N}\left(\int_1^x (\rho_1(\xi)+\rho_2(\xi)-n_d(\xi))\xi^{N-1}d\xi + c(\rho_1,\rho_2,t)\right), \qquad (3.3)$$

where $c(\rho_1,\rho_2,t)$ is given by the expression,

$$\frac{1}{\int_1^2 s^{1-N}ds}\left(\phi_2(t)-\phi_1(t)-\int_1^2 s^{1-N}\int_1^s(\rho_1(\xi)+\rho_2(\xi)-n_d(\xi))\xi^{N-1}d\xi ds\right).$$

We construct the approximate solutions $u_i^h(x,t)=(\rho_i^h(x,t),m_i^h(x,t))$ of (3.1) as in the construction for (2.19) with $u=u_i$, $G=G(u_1,u_2,x,t)$, for each $i=1,2$. Then (2.25) becomes

$$\begin{cases} \rho_i^h(x,t)=\rho_{i0}^h(x,t)+R_i(\rho_{10}^h(x,t),\rho_{20}^h(x,t))(t-n\Delta t),\\ m_i^h(x,t)=m_{i0}^h(x,t)+G_2(u_{10}^h(x,t),u_{20}^h(x,t),x,t)(t-n\Delta t), \end{cases} \qquad (3.4)$$

for $n\Delta t\le t<(n+1)\Delta t$. Next we make some estimates on the approximate solutions, and then prove the convergence of the approximate solutions.

For the coupling terms $R_i(\rho_1,\rho_2)$ and $H_i(\rho_1,\rho_2,E_1,E_2)$, we assume that (A1) and (A2) of Section 1.3 hold. For ease of reference, we repeat these here. Thus, we assume that R_i and H_i are Lipschitz continuous functions of the variables $\rho_1\ge 0$, $\rho_2\ge 0$, E_1 and E_2; and there exists a decomposition of R_i: $R_i=R_i^+-R_i^-$, with $R_i^\pm(\rho_1,\rho_2)>0$, and a constant $C>0$, such that, for all ρ_1, $\rho_2>0$, and each $i=1,2$,

$$R_i(\rho_1,\rho_2)\le C, \qquad (3.5)$$

$$\frac{R_i^+(\rho_1,\rho_2)}{\rho_i}\le C, \quad \text{if } R_i^+(\rho_1,\rho_2)\ge R_i^-(\rho_1,\rho_2) \text{ and } \rho_i\ge \left(\frac{\theta}{\theta+1}\right)^{\frac{1}{\theta}}, \qquad (3.6)$$

$$\frac{R_i^-(\rho_1,\rho_2)}{\rho_i}\le C, \quad \text{if } R_i^+(\rho_1,\rho_2)\le R_i^-(\rho_1,\rho_2), \qquad (3.7)$$

$$\frac{|H_i(\rho_1,\rho_2,E_1,E_2)|}{\rho_i}\le C. \qquad (3.8)$$

The above assumptions are quite general, as noted in the Introduction.

3.1 Uniform Estimates

In this section, we shall derive the L^∞ estimates of the approximate solution $u_i^h(x,t)=(\rho_i^h(x,t),m_i^h(x,t))$ for each $i=1,2$. For simplicity of notation, we will drop the index i of the approximate solution $u_i^h=(\rho_i^h,m_i^h)$ and $u_{i0}^h=(\rho_{i0}^h,m_{i0}^h)$ in the proofs, as well as those of R_i, and denote by $C>0$ a universal constant depending only on T, throughout this paper.

First, we have the following lemma about the conservation of particles:

Lemma 3.1 *If (3.5) holds, then there exists a constant $C > 0$ which depends only on the bounds of $A(x)$ and $\rho_{i0} + \left|\frac{m_{i0}}{\rho_{i0}}\right|$, such that, for any $t \in [0, T]$ and each $i = 1, 2$,*

$$\int_1^2 \rho_{i0}^h(x,t)dx \leq C + C \max_j \{\rho_{ij}^n\} h^\beta,$$

for some n with $t \in [n\Delta t, (n+1)\Delta t)$.

Proof By the construction of the approximate solutions, one has

$$\int_1^2 A(x)\rho_0^h(x,(n+1)\Delta t - 0)dx + \int_{n\Delta t}^{(n+1)\Delta t} A(x(t)) \sum (\sigma[\rho_0^h] - [m_0^h])dt$$

$$= \int_1^2 A(x)\rho_0^h(x, n\Delta t + 0)dx + O(h^\beta \Delta t).$$

Using the Rankine–Hugoniot condition, one has

$$\int_{n\Delta t}^{(n+1)\Delta t} A(x(t)) \sum (\sigma[\rho_0^h] - [m_0^h])dt = 0.$$

Then

$$\int_1^2 A(x)\rho_0^h(x, (n+1)\Delta t - 0)dx$$

$$= \int_1^2 A(x)(\rho_0^h(x, n\Delta t + 0) - \rho_0^h(x, n\Delta t - 0))dx$$

$$+ \int_1^2 A(x)\rho_0^h(x, n\Delta t - 0)dx + O(h^{\beta+1}).$$

By the construction and Lemma 2.4, one has

$$\frac{1}{h} \int_{1+(j-\frac{1}{2}h)}^{1+(j+\frac{1}{2}h)} A(x)(\rho^h(x, n\Delta t - 0) - \rho_j^n)dx = 0, \tag{3.9}$$

$$\frac{1}{h} \int_{1+(j-\frac{1}{2}h)}^{1+(j+\frac{1}{2}h)} A(x)(\rho_0^h(x, n\Delta t + 0) - \rho_j^n)dx = \rho_j^n O(h^{1+\beta}). \tag{3.10}$$

Then by (3.9)–(3.10), and (3.5):

$$\int_1^2 A(x)(\rho_0^h(x, n\Delta t + 0) - \rho_0^h(x, n\Delta t - 0))dx$$

$$= \sum_j \int_{1+(j-\frac{1}{2})h}^{1+(j+\frac{1}{2})h} A(x)(\rho_0^h(x, n\Delta t + 0) - \rho_j^n)dx$$

$$+ \sum_j \int_{1+(j-\frac{1}{2})h}^{1+(j+\frac{1}{2})h} A(x)(\rho_j^n - \rho_0^h(x, n\Delta t - 0))dx$$

$$= \sum_j \int_{1+(j-\frac{1}{2})h}^{1+(j+\frac{1}{2})h} A(x) R(\rho_{10}^h(x, n\Delta t - 0), \rho_{20}^h(x, n\Delta t - 0)) \Delta t dx$$
$$+ \sum_j \rho_j^n O(h^{2+\beta})$$
$$\leq \sum_j \rho_j^n O(h^{2+\beta}) + Ch,$$

and

$$\int_1^2 A(x) \rho_0^h(x, (n+1)\Delta t - 0) dx$$
$$\leq \int_1^2 A(x) \rho_0^h(x, n\Delta t - 0) dx + \sum_j \rho_j^n O(h^{2+\beta}) + O(h^{1+\beta}) + Ch.$$

Therefore, by induction on n, we have, for any positive integer n,

$$\int_1^2 A(x) \rho_0^h(x, n\Delta t - 0) dx \leq \int_1^2 A(x) \rho_0^h(x, 0) dx + \sum_j \rho_j^n O(h^{1+\beta}) + O(h^\beta) + C.$$

For any $t \in [0, T]$, $t \in [n\Delta t, (n+1)\Delta t)$ for some n, then, by the construction of the approximate solutions and the Rankine–Hugoniot condition, one has

$$\int_1^2 A(x) \rho_0^h(x, t) dx = \int_1^2 A(x) \rho_0^h(x, n\Delta t + 0) dx + O(h^\beta \Delta t)$$
$$\leq \int_1^2 A(x) \rho_0(x) dx + \sum_j \rho_j^n O(h^{1+\beta}) + O(h^\beta) + C.$$

Since $A(x) \geq N \omega_N$ for any $x \in (1, 2)$, Lemma 3.1 follows. □

Let $\Pi_T = [1, 2] \times [0, T]$. We have the following uniform estimate.

Theorem 3.2 *Suppose that (3.5)–(3.8) hold and there exists a constant $C > 0$ such that $0 < \rho_{i0}(x) \leq C$, and $|v_{i0}(x)| \leq C$ for all $x \in (1, 2)$, $i = 1, 2$. Then, for $h \leq h_0$, there exists a positive constant $C(T)$, independent of h and τ_i, such that, for each $i = 1, 2$,*

$$h^\beta \leq \rho_i^h(x, t) \leq C(T), \quad |v_i^h(x, t)| \leq C(T), \quad (x, t) \in \Pi_T.$$

Proof We again suppress the subscripts $i = 1, 2$ in the course of the proof. Suppose, for small h,

$$\sup_{(x,t)} \rho^h(x, t) \leq \frac{1}{h^\beta}, \quad \text{and} \quad \sup_{(x,t)} |v_0^h(x, t)| \leq \frac{1}{h^\beta},$$

and there exists $K(h)$ satisfying $K(h) \to \infty$ as $h \to 0$ such that
$$\sup_{(x,t)} \rho^h(x,t) \leq K(h),$$
and then
$$\sup_{(x,t)} |R(\rho_{10}^h(x,t), \rho_{20}^h(x,t))| \leq \frac{1}{h^\beta}.$$
Thus, Lemma 3.1 implies
$$\int_1^2 \rho_0^h(x,t)dx \leq C, \quad \forall\, t \in [0,T].$$
By the construction of (ρ^h, m^h), we have $\rho_0^h(x,t) \geq 2h^\beta$, for $(x,t) \in \Pi_T$. Then for $t \in [n\Delta t, (n+1)\Delta t)$, by (3.4),
$$\rho^h(x,t) \geq 2h^\beta - \frac{Ch}{h^\beta} \geq h^\beta.$$
By (3.3), one has
$$|\phi_x| \leq C.$$
In $n\Delta t \leq t < (n+1)\Delta t$, we estimate the Riemann invariant w using (3.4) and (3.1). Note that
$$\left|\frac{R}{\rho_0^h}(t - n\Delta t)\right| \leq \frac{1}{h^\beta} \cdot \frac{Ch}{2h^\beta} \leq Ch^{1-2\beta}.$$
Then, by by the definition of the Riemann invariants, by (3.4), (3.1) and (3.8), we have the following estimates:

$w(u^h(x,t))$
$$= \left(\frac{m_0^h}{\rho_0^h} + \left(\phi_x - \frac{m_0^h}{\tau\rho_0^h} + \frac{H}{\rho_0^h}\right)(t - n\Delta t)\right)\left(1 + \frac{R}{\rho_0^h}(t - n\Delta t)\right)^{-1}$$
$$+ \frac{(\rho_0^h)^\theta}{\theta}\left(1 + \frac{R}{\rho_0^h}(t - n\Delta t)\right)^\theta$$
$$= \left(\frac{m_0^h}{\rho_0^h} + \left(\phi_x - \frac{m_0^h}{\tau\rho_0^h} + \frac{H}{\rho_0^h}\right)(t - n\Delta t)\right)\left(1 - \frac{R}{\rho_0^h}(t - n\Delta t) + O(h^{2-4\beta})\right)$$
$$+ \frac{(\rho_0^h)^\theta}{\theta}\left(1 + \theta\frac{R}{\rho_0^h}(t - n\Delta t) + O(h^{2-4\beta})\right)$$
$$\leq \left(w(u_0^h(x,t)) + \left(\phi_x + \frac{H}{\rho_0^h} - \frac{m_0^h}{\tau\rho_0^h}\right)(t - n\Delta t)\right)(1 + O(\Delta t)) + C\Delta t$$
$$+ \left(\phi_x + \frac{H}{\rho_0^h} - \frac{m_0^h}{\tau\rho_0^h}\right)(t - n\Delta t)\left(-\frac{R}{\rho_0^h}(t - n\Delta t)\right)$$
$$- (u_0^h - (\rho_0^h)^\theta)\frac{R}{\rho_0^h}(t - n\Delta t)$$

$$\leq \left(w(u_0^h(x,t)) \left(1 - \frac{t-n\Delta t}{2\tau}\right) - z(u_0^h(x,t)) \frac{t-n\Delta t}{2\tau} \right) (1+O(\Delta t))$$
$$+ C\Delta t + (t-n\Delta t) I_1,$$

where
$$I_1 = -(v_0^h - (\rho_0^h)^\theta) \frac{R}{\rho_0^h}.$$

Similarly,
$$z(u^h(x,t)) \geq \left(z(u_0^h(x,t)) \left(1 - \frac{t-n\Delta t}{2\tau}\right) - w(u_0^h(x,t)) \frac{t-n\Delta t}{2\tau} \right) (1+O(\Delta t))$$
$$- C\Delta t + (t-n\Delta t) I_2,$$

where
$$I_2 = -(v_0^h + (\rho_0^h)^\theta) \frac{R}{\rho_0^h}.$$

It suffices to consider the cases $w(u_0^h(x,t)) \geq 1$ and $z(u_0^h(x,t)) \leq -1$. When $w(u_0^h(x,t)) \geq 1$,
$$v_0^h \geq 1 - \frac{(\rho_0^h)^\theta}{\theta}.$$

If $R \geq 0$, then
$$I_1 \leq -\left(1 - \frac{\theta+1}{\theta}(\rho_0^h)^\theta\right) \frac{R}{\rho_0^h};$$

in the case $(\rho_0^h)^\theta \leq \frac{\theta}{\theta+1}$, $I_1 \leq 0$; in the case $(\rho_0^h)^\theta \geq \frac{\theta}{\theta+1}$, by (3.6),
$$I_1 \leq \frac{\theta+1}{\theta}(\rho_0^h)^\theta \frac{R^+}{\rho_0^h} \leq C(\rho_0^h)^\theta \leq C(w(u_0^h) - z(u_0^h)).$$

If $R \leq 0$, then
$$I_1 = -(v_0^h - (\rho_0^h)^\theta) \frac{R}{\rho_0^h};$$

in the case $v_0^h \leq (\rho_0^h)^\theta$, $I_1 \leq 0$; in the case $v_0^h \geq (\rho_0^h)^\theta$, by (3.7),
$$I_1 \leq (v_0^h - (\rho_0^h)^\theta) \frac{R^-}{\rho_0^h} \leq C v_0^h \leq C w(u_0^h).$$

Similarly, when $z(u_0^h(x,t)) \leq -1$, $I_2 \geq 0$, or $I_2 \geq -C(z(u_0^h) - w(u_0^h))$, or $I_2 \geq Cz(u_0^h)$.

By Lemma 2.4 and the construction of (ρ^h, m^h), we have
$$\begin{cases} w(u_0^h(x,t)) \leq \max(\sup_x w(u_0^h(x,n\Delta t+0)), 1)(1+C\Delta t), \\ z(u_0^h(x,t)) \geq \min(\inf_x z(u_0^h(x,n\Delta t+0)), -1)(1+C\Delta t), \end{cases}$$

for $h \leq h_0$. Then

$$w(u^h(x,t)) \leq \max(\sup_x w(u_0^h(x, n\Delta t + 0)), 1)(1 + C\Delta t)\left(1 - \frac{t - n\Delta t}{2\tau}\right)$$
$$- \min(\inf_x z(u_0^h(x, n\Delta t + 0)), -1)(1 + C\Delta t)\frac{t - n\Delta t}{2\tau} + C\Delta t.$$

Similarly, we have

$$z(u^h(x,t)) \geq \min(\inf_x z(u_0^h(x, n\Delta t + 0)), -1)(1 + C\Delta t)\left(1 - \frac{t - n\Delta t}{2\tau}\right)$$
$$- \max(\sup_x w(u_0^h(x, n\Delta t + 0)), 1)(1 + C\Delta t)\frac{t - n\Delta t}{2\tau} + C\Delta t.$$

Note that
$$\begin{cases} w(u_j^n) = w(\bar{u}_j^n)(1 + O(h)), \\ z(u_j^n) = z(\bar{u}_j^n)(1 + O(h)), \end{cases}$$

with $\bar{u}_j^n = \frac{1}{h} \int_{1+(j-\frac{1}{2})h}^{1+(j+\frac{1}{2})h} u^h(x, n\Delta t - 0) dx$.

Set $M_n = \max(\sup_x w(u_0^h(x, n\Delta t + 0)), -\inf_x z(u_0^h(x, n\Delta t + 0)), 1)$. Then

$$M_{n+1} \leq M_n(1 + C\Delta t) + C\Delta t.$$

Thus,

$$\begin{aligned} M_{n+1} &\leq M_0(1 + C\Delta t)^{n+1} + C(n+1)\Delta t(1 + C\Delta t)^n \\ &\leq M_0(1 + C\Delta t)^{\frac{T}{\Delta t}+1} + CT(1 + C\Delta t)^{\frac{T}{\Delta t}} \leq C(T). \end{aligned}$$

This means:
$$\begin{cases} w(u^h(x,t)) \leq C(T)(1 + C\Delta t) + K\Delta t \leq C(T), \\ -z(u^h(x,t)) \leq C(T)(1 + C\Delta t) + K\Delta t \leq C(T), \\ w(u^h(x,t)) - z(u^h(x,t)) \geq \frac{2h^{\beta\theta}}{\theta}. \end{cases}$$

Therefore, there exists a constant $C(T) > 0$ such that

$$h^\beta \leq \rho^h(x,t) \leq C(T), \quad |v^h(x,t)| = \left|\frac{m^h(x,t)}{\rho^h(x,t)}\right| \leq C(T),$$

where the constant $C(T)$ is independent of h and τ.

Choose $h_0 > 0$ such that, for $h \leq h_0$, $C(T) < \min(\frac{1}{h^\beta}, K(h))$; then

$$h^\beta \leq \rho^h(x,t) \leq C(T) < \frac{1}{h^\beta}, \quad |v^h(x,t)| \leq C(T) < K(h).$$

This completes the proof. □

3.2 H^{-1} Compactness of Entropy Measures

We need the following basic lemma (cf. [3, 11, 29]) to prove the H^{-1} compactness of entropy measures for the approximate solutions (ρ_i^h, m_i^h).

Lemma 3.3 *Let $\Omega \subset \mathbb{R}^N$ be a bounded domain. Then*

$$\text{(compact set of } W^{-1,q}(\Omega)) \cap \text{(bounded set of } W^{-1,r}(\Omega))$$
$$\subset \text{(compact set of } W^{-1,2}_{\text{loc}}(\Omega)),$$

where q and r are constants, $1 < q \leq 2 < r < \infty$.

Theorem 3.4 *If (3.5)–(3.8) hold, and $\{u_i^h\}$, $i = 1, 2$, are the approximate solutions, then the measure sequence*

$$\eta(u_i^h)_t + q(u_i^h)_x$$

is a compact subset of $H^{-1}_{\text{loc}}(\Omega)$ for all weak entropy pairs (η, q), where Ω is any bounded and open set in Π_T.

Proof We drop the index i of u_i^h and u_{i0}^h in the proof. For any test function $\psi \in C_0^1(\Pi_T)$, we have

$$\iint_{\Pi_T} (\eta(u^h)\psi_t + q(u^h)\psi_x)dxdt = A(\psi) + M(\psi) + N(\psi) + L(\psi) + \Sigma(\psi) + E(\psi). \tag{3.11}$$

Here,

$$A(\psi) = \iint_{\Pi_T} \left((\eta(u^h) - \eta(u_0^h))\psi_t + (q(u^h) - q(u_0^h))\psi_x\right) dxdt,$$

$$M(\psi) = \int_1^2 \psi(x,T)\eta(u_0^h(x,T))dx - \int_1^2 \psi(x,0)\eta(u_0^h(x,0))dx,$$

$$N(\psi) = -\iint_{\Pi_T} a(x)g(u_0^h)\nabla\eta(u_0^h)\psi(x,t)dxdt,$$

$$\Sigma(\psi) = \int_0^T \sum (\sigma[\eta] - [q])\, \psi(x(t),t)dt,$$

$$L(\psi) = \sum_{j,n} \int_{1+(j-\frac{1}{2})h}^{1+(j+\frac{1}{2})h} (\eta(u_{0-}^n) - \eta(u_{0+}^n))\, \psi(x, n\Delta t)dx \equiv L_1(\psi) + L_2(\psi),$$

$$L_1(\psi) = \sum_{j,n} \psi_j^n \int_{1+(j-\frac{1}{2})h}^{1+(j+\frac{1}{2})h} (\eta(u_{0-}^n) - \eta(u_{0+}^n))dx,$$

$$L_2(\psi) = \sum_{j,n} \int_{1+(j-\frac{1}{2})h}^{1+(j+\frac{1}{2})h} (\eta(u_{0-}^n) - \eta(u_{0+}^n))(\psi - \psi_j^n)dx,$$

and

$$|E(\psi)| \leq Ch^\beta \|\psi\|_{H^1},$$

where $u_{0\pm}^n = u_0^h(x, n\Delta t \pm 0)$, $\psi_j^n = \psi(1+jh, n\Delta t)$, the summation in $\Sigma(\psi)$ is taken over all discontinuities in u_0^h at a fixed time t, σ is the propagating speed of the discontinuity, and $E(\psi)$ is the error term including the error in the steady-state solutions and the error near the vacuum in the construction of approximate solutions, and

$$[\eta] = \eta(u_0^h(x(t)+0,t)) - \eta(u_0^h(x(t)-0,t)),$$

$$[q] = q(u_0^h(x(t)+0,t)) - q(u_0^h(x(t)-0,t)),$$

are the jumps of $\eta(u_0^h(x,t))$ and $q(u_0^h(x,t))$ across a discontinuity $S = (x(t), t)$ in $u_0^h(x,t)$.

We shall make use of the following two lemmas:

Lemma 3.5 *For any n and $h \leq h_0$,*

$$\frac{1}{h}\int_{1+(j-\frac{1}{2})h}^{1+(j+\frac{1}{2})h} (\rho^h(x, n\Delta t - 0) - \rho_j^n)dx = O(h).$$

This follows immediately from (3.9).

Lemma 3.6 *There exists a constant $C > 0$ such that*

(1). $\displaystyle\sum_{j,n} \int_{1+(j-\frac{1}{2})h}^{1+(j+\frac{1}{2})h} \int_0^1 \Theta_\pm^n(\eta_*, s)dsdx \leq C$, *where*

$$\Theta_\pm^n(\eta, s) = (1-s)(u_{0\pm}^n - u_j^n)^\top \nabla^2\eta(u_j^n + s(u_{0\pm}^n - u_j^n))(u_{0\pm}^n - u_j^n).$$

(2). $\displaystyle\sum_{j,n} \int_{1+(j-\frac{1}{2})h}^{1+(j+\frac{1}{2})h} |u_{0\pm}^n - u_j^n|^2 dx \leq C.$

This lemma follows from a similar argument in [8].
We now use these lemmas to prove Theorem 3.4.
(a) From the lemmas in Section 2.1, Section 2.2, Lemma 3.5, and Lemma 3.6,

$$|M(\psi)| \leq \|\psi\|_{C_0(\Omega)} \int_1^2 (|\eta(u_0^h(x,T))| + |\eta(u_0^h(x,0))|)\,dx \leq C\|\psi\|_{C_0(\Omega)},$$

$$|N(\psi)| \leq \|\psi\|_{C_0(\Omega)} \|\nabla\eta\|_\infty \|a(x)g(u_0^h)\|_\infty T \leq C\|\psi\|_{C_0(\Omega)},$$

$$|\Sigma(\psi)| \leq \|\psi\|_{C_0(\Omega)} \int_0^T \left(\sum(\sigma[\eta_*] - [q_*]) + h^{2(1-\beta)}\right)dx \leq C\|\psi\|_{C_0(\Omega)},$$

and

$$|L_1(\psi)| \leq \left|\sum_{j,n} \psi_j^n \int_{1+(j-\frac{1}{2})h}^{1+(j+\frac{1}{2})h} (\eta(u_{0-}^n) - \eta(u_j^n))dx\right|$$

$$+\left|\sum_{j,n}\psi_j^n\int_{1+(j-\frac{1}{2})h}^{1+(j+\frac{1}{2})h}(\eta(u_{0+}^n)-\eta(u_j^n))dx\right|$$

$$\leq C\|\psi\|_{C_0}\sum_{j,n}\int_{1+(j-\frac{1}{2})h}^{1+(j+\frac{1}{2})h}\int_0^1\Theta_-^n(\eta_*,s)dsdx$$

$$+C\|\psi\|_{C_0}\sum_{j,n}\int_{1+(j-\frac{1}{2})h}^{1+(j+\frac{1}{2})h}\int_0^1\Theta_+^n(\eta_*,s)dsdx$$

$$+\|\psi\|_{C_0}(O(1)+O(h^{1-2\beta}))$$

$$\leq C\|\psi\|_{C_0}.$$

Hence
$$|(M+N+L_1+\Sigma)(\psi)|\leq C\|\psi\|_{C_0},$$
that is,
$$\|M+N+L_1+\Sigma\|_{C_0^*}\leq C.$$
By the embedding theorem, $(C_0(\Omega))^* \hookrightarrow W^{-1,q_1}$, for $1<q_1<2$,
$$M+N+L_1+\Sigma \quad \text{is compact in} \quad W^{-1,q_1}(\Omega).$$

(b) For any $\psi \in C_0^\alpha(\Omega)$, $\frac{1}{2}<\alpha<1$, using Hölder's inequality, we have

$$|L_2(\psi)|$$
$$\leq \sum_{j,n}\int_{1+(j-\frac{1}{2})h}^{1+(j+\frac{1}{2})h}|\psi-\psi_j^n|\left(|\eta(u_{0-}^n)-\eta(u_j^n)|+|\eta(u_{0+}^n)-\eta(u_j^n)|\right)dx$$
$$\leq C\|\psi\|_{C_0^\alpha}h^{\alpha-\frac{1}{2}}\|\nabla\eta\|_\infty$$
$$\times\left(\sum_{j,n}\int_{1+(j-\frac{1}{2})h}^{1+(j+\frac{1}{2})h}|u_{0-}^n-u_j^n|^2dx+\sum_{j,n}\int_{1+(j-\frac{1}{2})h}^{1+(j+\frac{1}{2})h}|u_{0+}^n-u_j^n|^2dx\right)^{\frac{1}{2}}$$
$$\leq Ch^{\alpha-\frac{1}{2}}\|\psi\|_{C_0^\alpha}.$$

By the Sobolev theorem: $W_0^{1,p}(\Omega) \subset C_0^\alpha(\Omega)$, $0<\alpha<1-\frac{2}{p}$, we have

$$|L_2(\psi)|\leq Ch^{\alpha-\frac{1}{2}}\|\psi\|_{W_0^{1,p}(\Omega)},\quad p>\frac{2}{1-\alpha},$$

that is,
$$\|L_2\|_{W^{-1,q_2(\Omega)}}\leq Ch^{\alpha-\frac{1}{2}}\to 0,\quad h\to 0,$$

for $1<q_2<\frac{2}{1+\alpha}$.

Therefore, L_2 is compact in W^{-1,q_2}. Then,

$$M+N+L+\Sigma \quad \text{is compact in} W^{-1,q_0},$$

where $1 < q_0 = \min(q_1, q_2) < \frac{2}{1+\alpha}$.

The uniform boundedness of the approximate solutions implies

$$M + N + L + \Sigma \text{ is bounded in } W^{-1,r}, r > 1.$$

By Lemma 3.3, $M + N + L + \Sigma$ is compact in H^{-1}_{loc}.

(c) Finally, for $A(\psi)$ we have

$$\|A(\psi)\| \leq \int\int_{\Pi_T} (\|\nabla \eta\|_\infty + \|\nabla q\|_\infty)(|\psi_t| + |\psi_x|)|u^h - u^h_0| dx dt \leq Ch \|\psi\|_{H^1_0(\Omega)}.$$

Since $C_0^\infty(\Omega)$ is dense in $H^1_0(\Omega)$, then

$$\|A\|_{H^{-1}_{\text{loc}}(\Omega)} \leq Ch \to 0, \quad \text{as} \quad h \to 0,$$

so A is compact in $H^{-1}_{\text{loc}}(\Omega)$.

By $\|E\|_{H^{-1}} \leq Ch^\beta \to 0$, as $h \to 0$, we know that E is compact in $H^{-1}_{\text{loc}}(\Omega)$. Therefore $A + M + N + L + \Sigma + E$ is compact in $H^{-1}_{\text{loc}}(\Omega)$, which means that $\eta(u^h)_t + q(u^h)_x$ is compact in $H^{-1}_{\text{loc}}(\Omega)$. □

3.3 Convergence and Existence

In this section, we prove that (3.1) has a weak entropy solution, which is the limit function of the approximate solutions.

Definition 3.7 *The measurable functions $u_i(x,t) = (\rho_i(x,t), m_i(x,t))$, $i = 1, 2$, are weak entropy solutions of (3.1) if, for any test function $\psi \in C_0^1(\Pi_T)$ with $\psi(1,t) = \psi(2,t) = \psi(x,T) = 0$ and, for each $i = 1, 2$,*

$$\int\int_{\Pi_T} (u_i \psi_t + f(u_i)\psi_x + (a(x)g(u_i) + G(u_1, u_2, x, t))\psi) \, dx dt \qquad (3.12)$$
$$+ \int_1^2 u_{i0}(x)\psi(x,0) dx = 0,$$

and, along any shock discontinuity with left state u_{i-}, right state u_{i+}, and speed σ_i,

$$\sigma_i(\eta(u_{i+}) - \eta(u_{i-})) - (q(u_{i+}) - q(u_{i-})) \geq 0, \qquad (3.13)$$

for any convex weak entropy pair (η, q).

Now we introduce the following compensated compactness framework (see [2, 3]):

Lemma 3.8 *Assume that the approximate solutions $u^h = (\rho^h, m^h)$ satisfy*

(1) There is a constant $C > 0$ such that $0 \leq \rho^h(x,t) \leq C$, $\left|\frac{m^h(x,t)}{\rho^h(x,t)}\right| \leq C$.

(2) The measure $\eta(u^h)_t + q(u^h)_x$ is compact in $H^{-1}_{\text{loc}}(\Omega)$, for all weak entropy pairs (η, q), where $\Omega \subset \Pi_T$ is any bounded and open set.

Then, for $1 < \gamma \leq 5/3$, there exists a convergent subsequence (still labeled u^h) such that $u^h(x,t) \longrightarrow u(x,t) = (\rho(x,t), m(x,t))$, a.e.

In Sections 3.1 and 3.2, we have proved that the approximate solutions $u_i^h(x,t)$, constructed at the beginning of Section 3 for (3.1), satisfy (1) and (2) of Lemma 3.8. Thus, we have the following theorem:

Theorem 3.9 *Suppose (3.5)–(3.8) hold. Then there is a convergent subsequence (still labeled u_i^h) of the approximate solutions $u_i^h(x,t) = (\rho_i^h(x,t), m_i^h(x,t))$, $i = 1, 2$, such that*

$$u_i^h(x,t) \longrightarrow u_i(x,t) = (\rho_i(x,t), m_i(x,t)), \text{ a.e.} \quad \text{as} \quad h \to 0,$$

and the function $u_i(x,t)$ is a weak entropy solution of (3.1) in the sense of Definition 3.7 and satisfies

$$0 \le \rho_i(x,t) \le C(T), \qquad \left|\frac{m_i(x,t)}{\rho_i(x,t)}\right| \le C(T),$$

for $(x,t) \in \Pi_T$, where $C(T) > 0$ is a constant.

Proof Again we drop the index i of u_i^h and of u_{i0}^h in the proof. We also drop the corresponding index for R_i. We now prove that $u(x,t)$ satisfies (3.12) and (3.13). Let $\psi \in C_0^1(\Pi_T)$ be any test function with $\psi(1,t) = \psi(2,t) = \psi(x,T) = 0$. Set $\bar{\psi}(x,t) = \frac{\psi(x,t)}{A(x)} \in C_0^1(\Pi_T)$. Then

$$\int\int_{\Pi_T} (\rho^h \psi_t + m^h \psi_x + (a(x)m^h + R(\rho_1^h, \rho_2^h))\psi)dxdt + \int_1^2 \rho_0^h(x)\psi(x,0)dx$$

$$= \int\int_{\Pi_T} (A(x)\rho^h \bar{\psi}_t + A(x)m^h \bar{\psi}_x + A(x)R(\rho_1^h, \rho_2^h)\bar{\psi})dxdt$$

$$+ \int_1^2 A(x)\rho_0^h(x)\bar{\psi}(x,0)dx$$

$$= \int\int_{\Pi_T} A(x)((\rho^h - \rho_0^h)\bar{\psi}_t + (m^h - m_0^h)\bar{\psi}_x)dxdt$$

$$+ \int\int_{\Pi_T} A(x)(R(\rho_1^h, \rho_2^h) - R(\rho_{10}^h, \rho_{20}^h))\bar{\psi}dxdt$$

$$+ \int_0^T \sum A(x(t))(\sigma[\rho_0^h] - [m_0^h])\bar{\psi}(x(t),t)dt + I_{11} + I_{12} + O(h^\beta)\|\bar{\psi}\|_{H^1},$$

where

$$I_{11} = \sum_{j,n} \bar{\psi}_j^n \int_{1+(j-\frac{1}{2})h}^{1+(j+\frac{1}{2})h} A(x)(\rho_0^h(x,n\Delta t - 0) - \rho_0^h(x,n\Delta t + 0))dx$$

$$+ \int\int_{\Pi_T} A(x)R(\rho_{10}^h, \rho_{20}^h)\bar{\psi}dxdt,$$

$$I_{12} = \sum_{j,n} \int_{1+(j-\frac{1}{2})h}^{1+(j+\frac{1}{2})h} A(x)(\bar{\psi} - \bar{\psi}_j^n)(\rho_0^h(x,n\Delta t - 0) - \rho_0^h(x,n\Delta t + 0))dx.$$

By the fractional step procedure and the Rankine–Hugoniot condition, one has

$$\left|\iint_{\Pi_T} A(x)((\rho^h - \rho_0^h)\bar{\psi}_t + (m^h - m_0^h)\bar{\psi}_x)dxdt\right| \leq O(h)\|\bar{\psi}\|_{C_0^1},$$

$$\left|\iint_{\Pi_T} A(x)(R(\rho_1^h, \rho_2^h) - R(\rho_{10}^h, \rho_{20}^h))\bar{\psi}dxdt\right| \leq O(h)\|\bar{\psi}\|_{C_0^1},$$

$$\int_0^T \sum A(x(t))(\sigma[\rho_0^h] - [m_0^h])\bar{\psi}(x(t), t)dt = 0.$$

From Proposition 3 in [25] or Lemma 2.4 in [4], and our Lemma 3.6, we have

$$\sum_{j,n} \int_{(n-1)\Delta t}^{n\Delta t} \int_{1+(j-\frac{1}{2})h}^{1+(j+\frac{1}{2})h} \left|\rho_{i0}^h(x, t) - \rho_{i0}^h(x, n\Delta t - 0)\right| dxdt \leq C(T)h, \quad i = 1, 2,$$

for some constant $C(T)$. Then by the construction of $u^h(x,t)$ and Lemma 2.4, one has

$$|I_{11}|$$

$$\leq \left|\sum_{j,n} \bar{\psi}_j^n \int_{1+(j-\frac{1}{2})h}^{1+(j+\frac{1}{2})h} (A(x)(\rho^h(x, n\Delta t - 0) - \rho_j^n)\right.$$

$$\left. + A(x)(\rho_j^n - \rho_0^h(x, n\Delta t + 0)))dx\right|$$

$$+ \left|\sum_{j,n} \int_{(n-1)\Delta t}^{n\Delta t} \int_{1+(j-\frac{1}{2})h}^{1+(j+\frac{1}{2})h} A(x)(R(\rho_{10}^h(x,t), \rho_{20}^h(x,t))\bar{\psi}(x,t)\right.$$

$$\left. - R(\rho_{10}^h(x, n\Delta t - 0), \rho_{20}^h(x, n\Delta t - 0))\bar{\psi}_j^n)dxdt\right|$$

$$\leq \|\bar{\psi}\|_{C_0^1} O(h^\beta) + \sum_{j,n} \int_{(n-1)\Delta t}^{n\Delta t} \int_{1+(j-\frac{1}{2})h}^{1+(j+\frac{1}{2})h} A(x) \left|(R(\rho_{10}^h(x,t), \rho_{20}^h(x,t))\right.$$

$$\left. - R(\rho_{10}^h(x, n\Delta t - 0), \rho_{20}^h(x, n\Delta t - 0)))\bar{\psi}(x,t)\right| dxdt$$

$$+ \sum_{j,n} \int_{(n-1)\Delta t}^{n\Delta t} \int_{1+(j-\frac{1}{2})h}^{1+(j+\frac{1}{2})h} \left|R_i(\rho_{10}^h(x, n\Delta t - 0), \rho_{20}^h(x, n\Delta t - 0))\right.$$

$$\left. \times (\bar{\psi}(x,t) - \bar{\psi}_j^n)\right| dxdt$$

$$\leq \|\bar{\psi}\|_{C_0^1} \sum_{i=1,2} \sum_{j,n} \int_{(n-1)\Delta t}^{n\Delta t} \int_{1+(j-\frac{1}{2})h}^{1+(j+\frac{1}{2})h} \left|\rho_{i0}^h(x,t) - \rho_{i0}^h(x, n\Delta t - 0)\right| dxdt$$

$$+ \|\bar{\psi}\|_{C_0^1}(O(h) + O(h^\beta))$$

$$\leq \|\bar{\psi}\|_{C_0^1}(O(h) + O(h^\beta)).$$

By Lemma 3.6,

$$|I_{12}| \leq C\|\bar{\psi}\|_{C_0^1}\sqrt{h}$$
$$\times \left(\sum_{j,n}\int_{1+(j-\frac{1}{2})h}^{1+(j+\frac{1}{2})h}(|\rho_0^h(x,n\Delta t-0)-\rho_j^n|^2+|\rho_j^n-\rho_0^h(x,n\Delta t+0)|^2)dx\right)^{\frac{1}{2}}$$
$$\leq O(\sqrt{h})\|\bar{\psi}\|_{C_0^1}.$$

Therefore

$$\iint_{\Pi_T}(\rho^h\psi_t + m^h\psi_x + a(x)m^h\psi)dxdt + \int_1^2 \rho_0^h(x)\psi(x,0)dx$$
$$= \|\bar{\psi}\|_{C_0^1}(O(h)+O(h^\beta)+O(\sqrt{h})) + \|\psi\|_{H^1}O(h^\beta) \to 0, \quad h \to 0.$$

Taking the limit $h \to 0$ on both sides and using the dominated convergence theorem, we have

$$\iint_{\Pi_T}(\rho\psi_t + m\psi_x + a(x)m\psi)dxdt + \int_1^2 \rho_0(x)\psi(x,0)dx = 0.$$

Following the similar estimates in [8], we have

$$\iint_{\Pi_T}\left(m\psi_t + \left(\frac{m^2}{\rho}+p(\rho)\right)\psi_x + \left(a(x)\frac{m^2}{\rho}+G_2(u_1,u_2,x,t)\right)\psi\right)dxdt$$
$$+ \int_1^2 m_0^h(x)\psi(x,0)dx = 0,$$

and, for any convex weak entropy pair (η,q), the limit function $u = (\rho,m)$ satisfies

$$\eta(u)_t + q(u)_x - (a(x)g(u) + G(u,x,t))\nabla\eta(u) \leq C,$$

in the sense of distributions. Using the standard procedure (cf. [21]), we conclude that the limit function $u(x,t)$ satisfies the entropy condition (3.13) along any shock wave. The uniform boundedness of $u^h(x,t)$ implies the boundedness of the weak solution $u(x,t)$.

The initial-boundary values can be recovered by the detailed estimates similar to [8] on the traces, which are properly defined along the boundaries. $\square$

3.4 Nozzle Solutions

Now we consider the following equations for the nozzle flow:

$$\begin{cases}(A\rho_i)_t + (Am_i)_x = AG_1, \\ (Am_i)_t + (A\frac{m_i^2}{\rho_i})_x + Ap(\rho_i)_x = AG_2, \quad i = 1,2,\end{cases} \qquad (3.14)$$

$1 < x < 2$, $t > 0$, with initial-boundary conditions:

$$\begin{cases} (\rho_i, m_i)|_{t=0} = (\rho_{i0}(x), m_{i0}(x)), \\ m_i|_{x=1} = m_i|_{x=2} = 0, \end{cases} \quad (3.15)$$

where $A(x) \in C^2$ represents the cross-sectional area of the nozzle at x, and G_1 and G_2 are the same as in (3.2).

The system (3.14) is equivalent to

$$\begin{cases} \partial_t \rho_i + \partial_x m_i = a(x) m_i + R_i(\rho_1, \rho_2), \\ \partial_t m_i + \partial_x \left(\frac{m_i^2}{\rho_i} + p(\rho_i) \right) = a(x) \frac{m_i^2}{\rho_i} + G_2(\rho_1, m_1, \rho_2, m_2, x, t), \quad i = 1, 2, \end{cases} \quad (3.16)$$

where $a(x) = -\frac{A'(x)}{A(x)}$.

As earlier in Section 3, we construct the approximate solutions $(\rho_i^h, m_i^h)(x, t)$ of (3.16) and then prove that the approximate solutions satisfy the compensated compactness framework (Lemma 3.8) as for (3.1). Then we conclude that there is a subsequence of the approximate solutions strongly convergent to the L^∞ function $(\rho_i(x,t), m_i(x,t))$ almost everywhere. We obtain:

Theorem 3.10 *Assume that the initial data (ρ_0, u_0) are bounded in L^∞. Then there exists a bounded weak entropy solution $(\rho_i(x,t), m_i(x,t))$ of (3.14–3.15) in the sense of Definition 3.7.*

4 The Simulation of the Gunn Diode

The numerical scheme we use in the simulations is the third-order ENO (Essentially Non-Oscillatory) scheme based on point values and Runge–Kutta time discretizations [28]. The description of this scheme applied to the hydrodynamic models can be found in [18], hence will not be repeated here.

The two-valley GaAs hydrodynamic model we use for this purpose, in one space dimension, has the following form, where j is to be selected distinct from i:

$$\begin{cases} \partial_t \rho_i + \partial_x(\rho_i v_i) = -\frac{\rho_i}{\tau_{nij}} + \frac{m_i^*}{m_j^*} \frac{\rho_j}{\tau_{nji}}, \\ \partial_t(\rho_i v_i) + \partial_x(\rho_i v_i^2 + p_i) = -\frac{e}{m_i^*} \rho_i F - \frac{\rho_i v_i}{\tau_{pi}}, \\ \partial_t E_i + \partial_x[v_i(E_i + p_i)] = -\frac{e}{m_i^*} \rho_i v_i F - \frac{E_i - \frac{3}{2}\frac{k_b}{m_i^*}\rho_i T_0}{\tau_{wii}} \\ \quad - \frac{E_i}{\tau_{wij}} + \frac{E_j}{\tau_{wji}} + \partial_x(\kappa_i \partial_x T_i), \end{cases} \quad (4.1)$$

for $i, j = 1, 2$ and $i \neq j$. Here, $\rho_i = m_i^* n_i$ are the particle densities, with n_i denoting the concentration, v_i are the particle velocities, E_i are the total energies, $p_i = (\gamma - 1)(E_i - \frac{1}{2}\rho_i v_i^2)$ are the pressures, with constant $\gamma = \frac{5}{3}$, $T_i = \frac{m_i^* p_i}{k_b \rho_i}$ are the temperatures. The equation (4.1) is coupled with the potential equation,

$$\epsilon \partial_x^2 \phi = e(n_1 + n_2 - n_d) \quad (4.2)$$

through the electric field term $F = -\partial_x \phi$. The relaxation terms are defined by

$$\frac{1}{\tau_{n12}} = \frac{1}{\tau_{p12}} = \begin{cases} 0, & \text{if } E_1 \leq \alpha n_1(1-\beta), \\ \text{smooth}, & \text{if } \alpha n_1(1-\beta) < E_1 < \alpha n_1(1+\beta), \\ 30, & \text{if } E_1 \geq \alpha n_1(1+\beta), \end{cases} \quad (4.3)$$

where α is a threshold controlling the amount of coupling between the valleys, and turns out to be crucial for the simulation. This will be discussed in more detail later. β is chosen to smooth the relaxation expressions. In our computation, $\beta = 0.15$ is used. The smooth connection between the two constant values 0 and 30 in (4.3) is achieved by a polynomial which makes the expression globally C^3. Other relaxation terms are defined by:

$$\frac{1}{\tau_{n21}} = \frac{1}{\tau_{p21}} = 2, \quad \frac{1}{\tau_{p11}} = 7, \quad \frac{1}{\tau_{p22}} = 20, \quad (4.4)$$

$$\frac{1}{\tau_{p1}} = \frac{1}{\tau_{p11}} + \frac{1}{\tau_{p12}}, \quad \frac{1}{\tau_{p2}} = \frac{1}{\tau_{p22}} + \frac{1}{\tau_{p21}}, \quad (4.5)$$

$$\tau_{w11} = 2\tau_{p11}, \quad \tau_{w22} = 2\tau_{p22}, \quad \tau_{w12} = 2\tau_{n12}, \quad \tau_{w21} = 2\tau_{n21}. \quad (4.6)$$

The heat conduction term κ_i is defined by

$$\kappa_i = \frac{3\mu_{0i} k_b^2 T_0}{2e} n_i \quad (4.7)$$

where $\mu_{0i} = \frac{e\tau_{pi}}{m_i^*}$. The values of other parameters used in the simulations (in our units) are: $m_1^* = 0.065 m_e^*$, $m_2^* = 0.222 m_e^*$ with $m_e^* = 0.9109$; $e = 0.1602$; $k_b = 0.138046 \times 10^{-4}$; $\epsilon = 12.9 \times 8.85418$; $T_0 = 300$. The doping n_d is defined by

$$n_d(x) = \begin{cases} 10^5, & 0 \leq x \leq 0.125 - \beta_1, \\ 10^4, & 0.125 + \beta_1 \leq x \leq 0.15 - \beta_1, \\ 5 \times 10^3, & 0.15 + \beta_1 \leq x \leq 0.1875 - \beta_1, \\ 10^4, & 0.1875 + \beta_1 \leq x \leq 1.875 - \beta_1, \\ 10^5, & 1.875 + \beta_1 \leq x \leq 2, \\ \text{smooth}, & \text{otherwise}, \end{cases} \quad (4.8)$$

where again "smooth" means a connection by a polynomial to make the doping globally C^3 and the smoothing length is chosen as $\beta_1 = 0.005$. We show the doping, in a logarithm scale, in Fig. 2. The fundamental units which we employ are the micron, the picosecond, the volt, and 10^{-30}kg for length, time, potential, and mass, respectively.

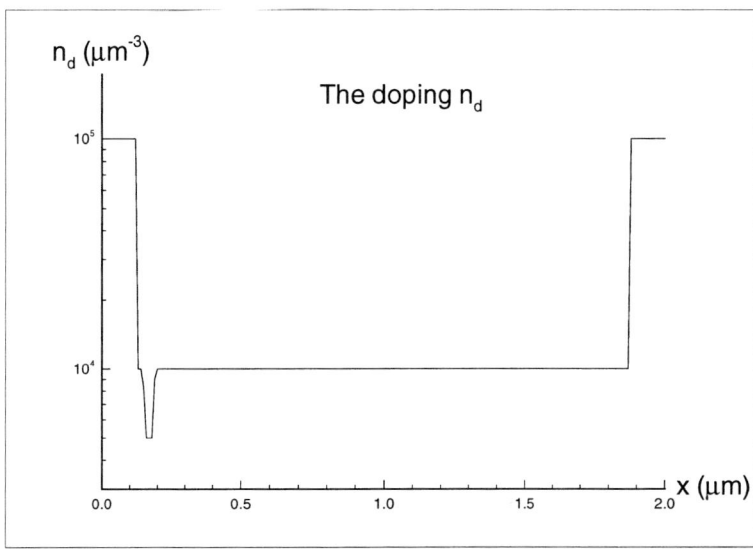

FIG. 2. One-dimensional two-valley hydrodynamic model. The doping n_d (in logarithm scale).

Boundary conditions are chosen as follows: for the concentrations n_i, we fix them at both ends with $n_1 = n_d$ and $n_2 \approx 0$. Technically, n_2 cannot be zero in the code, hence a small number $n_2 = 10^{-5}$ is used instead. The temperatures $\mathcal{T}_i$ are also fixed at both ends, with $\mathcal{T}_i = \mathcal{T}_0$. The velocity v_i satisfies a Neumann boundary condition (numerically it corresponds to zeroth-order extrapolation). The potential ϕ is fixed at both ends with a voltage difference determined by $vbias$ (the voltage bias): we take $\phi = 0$ at the left end $x = 0$, and $\phi = vbias = 2V$ at $x = 2$ for the stand-alone device, and $\phi = V_d(t)$ at $x = 2$, if the system is coupled with the Gunn oscillator (1.1).

For the stand-alone device, we would like to reach a steady-state solution of the system (4.1)–(4.2). We thus start from the following initial condition: $n_1(x,0) = n_d(x)$, $n_2(x,0) = 10^{-5}$, $v_i(x,0) = 0$, $\mathcal{T}_i(x,0) = \mathcal{T}_0$, and compute the time evolution of the system until it reaches a steady state. In practice, in order to achieve a steady state more rapidly, a continuation in $vbias$ is used, by starting from $vbias = 0.0$ V, and each time increasing it in increments of $0.05V$, by using the previous steady state as the initial condition for the higher value of $vbias$, until a steady state for the choice $vbias = 2$ V is reached. This steady state-solution is used as the initial condition for the Gunn oscillator, when the system (4.1)–(4.2) is coupled with the ODE (1.1).

It turns out that the coupling of the two valleys in the model, through an energy transfer, is crucially dependent upon the coupling threshold α in (4.3). The higher the threshold value, the lower the coupling effect becomes, as expected. Figure 3 shows the concentrations n_i of the two valleys in steady state, for different values of α.

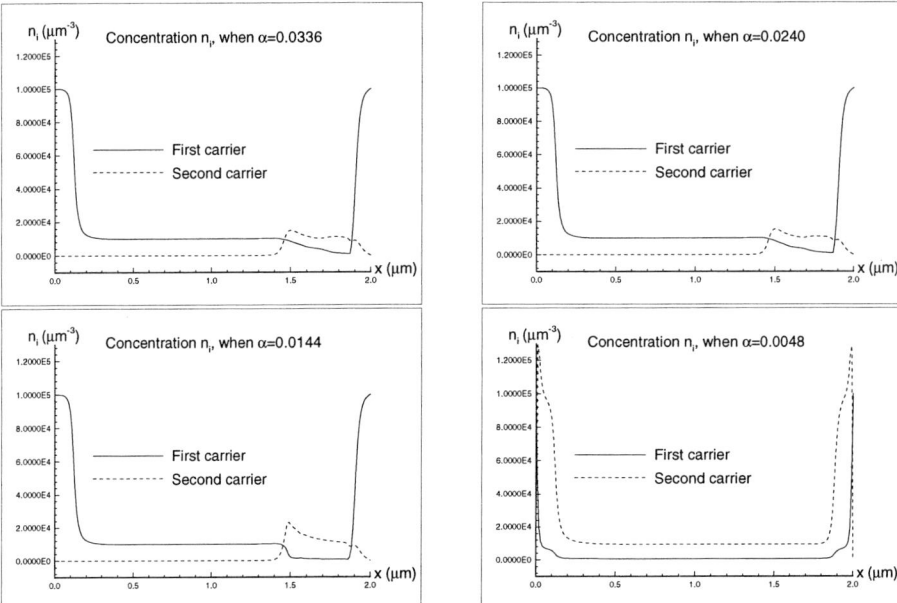

FIG. 3. One-dimensional two-valley hydrodynamic model. Concentrations n_i for both valleys at various coupling thresholds: $\alpha = 0.0336, 0.0240, 0.0144$, and 0.0048.

To conform with physics, and the Monte-Carlo simulation results in [22], we choose $\alpha = 0.0144$ for our simulation with the Gunn oscillator. The ODE (1.1) is coupled to the two valley hydrodynamic model (4.1)–(4.2), through the application of $V_D(t)$ as the boundary condition for the potential equation (4.2) at the right end $x = 2$. Other parameters in (1.1), expressed in values consistent with our units, are $V_B = vbias = 2$, $L = 3.5 \times 10^{-6}$, $C = \frac{\epsilon A}{W} + 0.82 \times 10^6$ with $A = 10^3$ and the device length $W = 2$, $R = 25 \times 10^{-6}$, and $I_d(t) = \frac{1}{W} \int_0^W I_e(x,t)dx$, with $I_e(x,t) = -eA\left(n_1(x,t)v_1(x,t) + n_2(x,t)v_2(x,t)\right)$. The initial condition for the ODE (1.1) is chosen as $I(0) = 0$ and $V_D(0) = 2$. Since we select 10^{-18} C and 10^{-18} F as basic units of charge and capacitance, the unit of inductance is 10^{-6} Henrys, and the unit of resistance is 10^6 Ohms. When the oscillator is coupled to the hydrodynamic system, after an initial transition, a time-periodic solution results. In Fig. 4, we show the time history of the applied voltage $V_D(t)$ (left) and the current flowing through the battery $I(t)$ (right) after the initial transition. These are the two variables in the ODE (1.1), and clearly they show sustained oscillations of a slow frequency layered over a fast frequency.

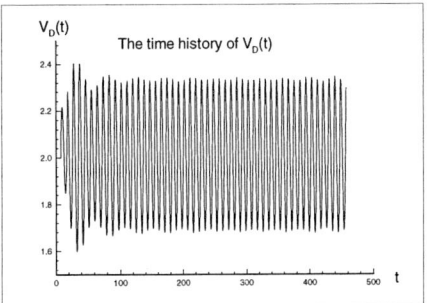
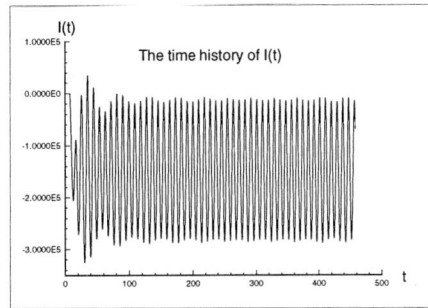

FIG. 4. One-dimensional two-valley hydrodynamic model, coupled with the Gunn oscillator. Left: the voltage at the device terminal $V_D(t)$; right: the current flowing through the battery $I(t)$.

We would like to point out that, since the simulation here involves a time-dependent system with strong hyperbolic components, which must be simulated for a very long time, upwinding and high order accuracy in space and time are important, justifying the usage of ENO schemes. Next, in Fig. 5, we show the concentration n_i, the velocity v_i, and the temperature $\mathcal{T}_i$ of both carriers, at 4 equally spaced "snaps" over one period of the oscillation. We can see the movement, or "precession", of the structure clearly in such a period.

5 The Simulation of the MESFET: Symmetry and Symmetry-Breaking

The motivation of this section is as follows: if some symmetry exists for the two-dimensional (2D) MESFET, which is relatively costly to simulate, we can use this information to reduce our model to 1D, at least for some components, thereby reducing the cost of simulation. For this purpose, we first look at the 2D simulation results at various values of *vbias*. The boundary condition at the gate is always taken as $vbiasgate = -0.4vbias$. The simulation is the same as those in [18], using a third-order ENO scheme and 192×64 grid points. The result is shown in Figs. 6–9.

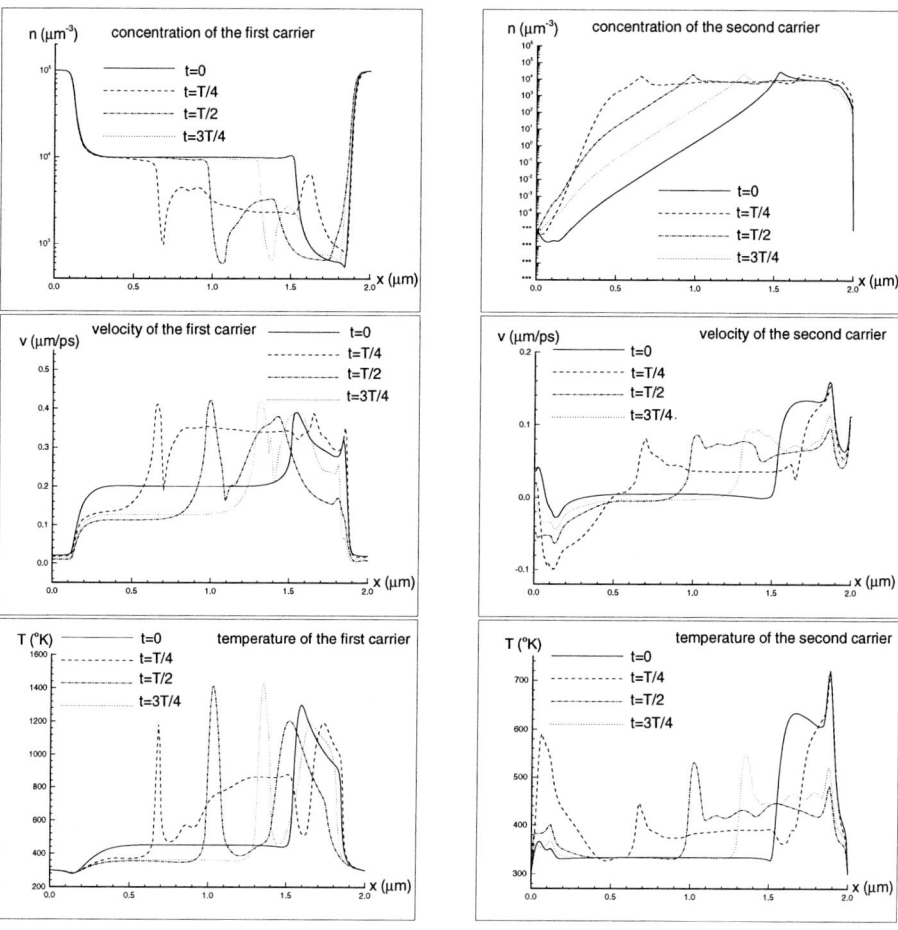

FIG. 5. One-dimensional two-valley hydrodynamic model, coupled with the Gunn oscillator. Four equally spaced "snaps" over one period of the oscillation. Top: concentration n_i; middle: velocity v_i; bottom: temperature $\mathcal{T}_i$. Left: the first valley $i = 1$; right: the second valley $i = 2$.

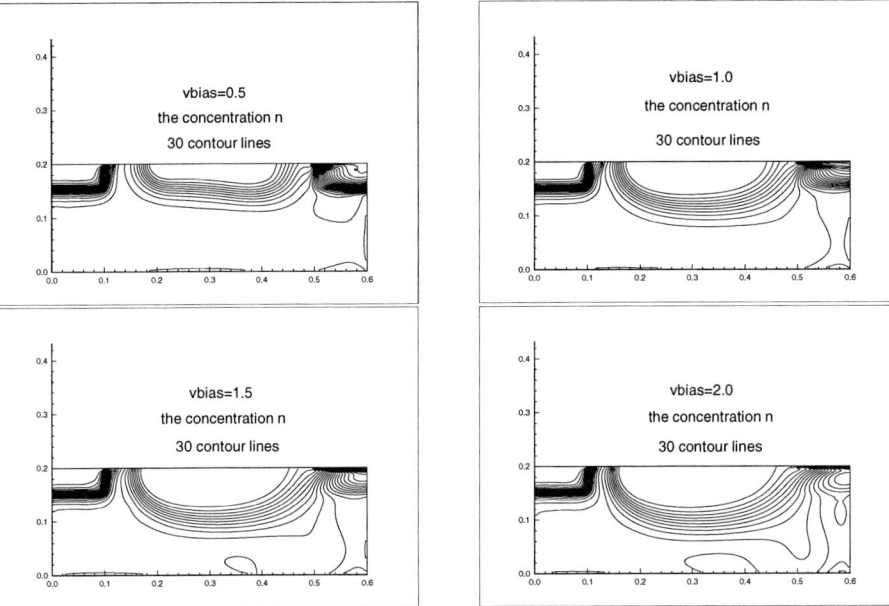

FIG. 6. Two-dimensional MESFET. Simulation result at $vbias = 0.5$ V, 1.0 V, 1.5 V and 2.0 V. The concentration n.

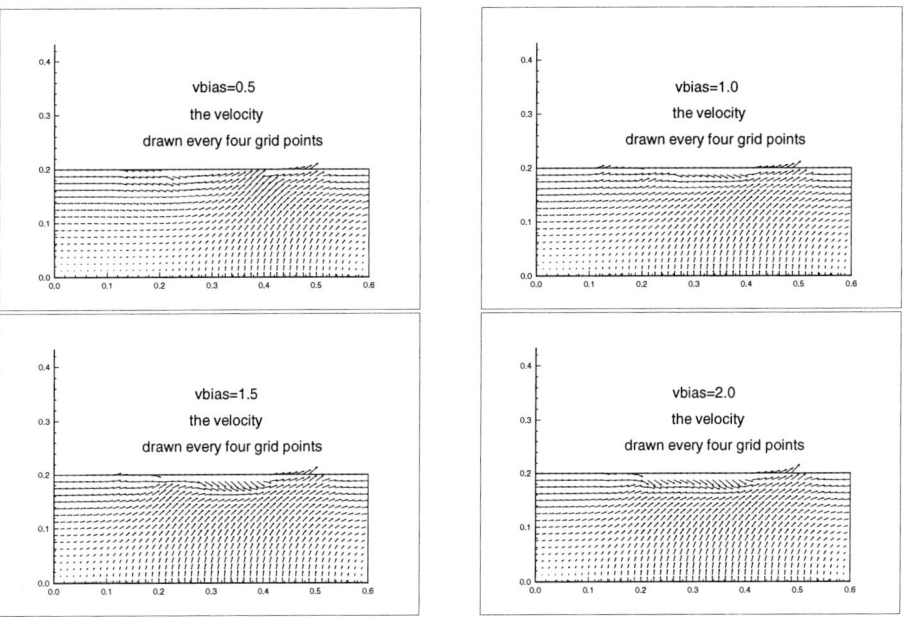

FIG. 7. Two-dimensional MESFET. Simulation result at $vbias = 0.5$ V, 1.0 V, 1.5 V and 2.0 V. The velocity vector $\vec{v}$.

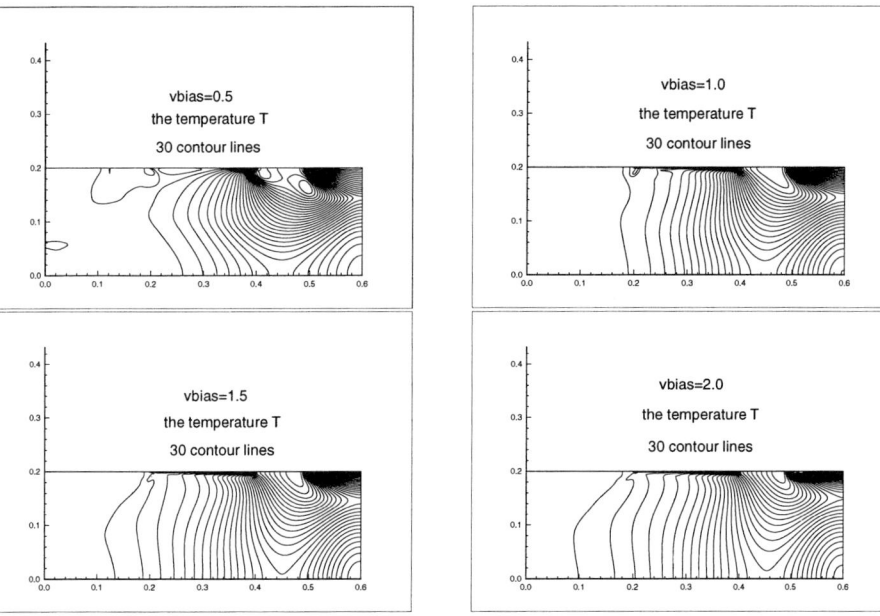

FIG. 8. Two-dimensional MESFET. Simulation result at $vbias = 0.5$ V, 1.0 V, 1.5 V and 2.0 V. The temperature T.

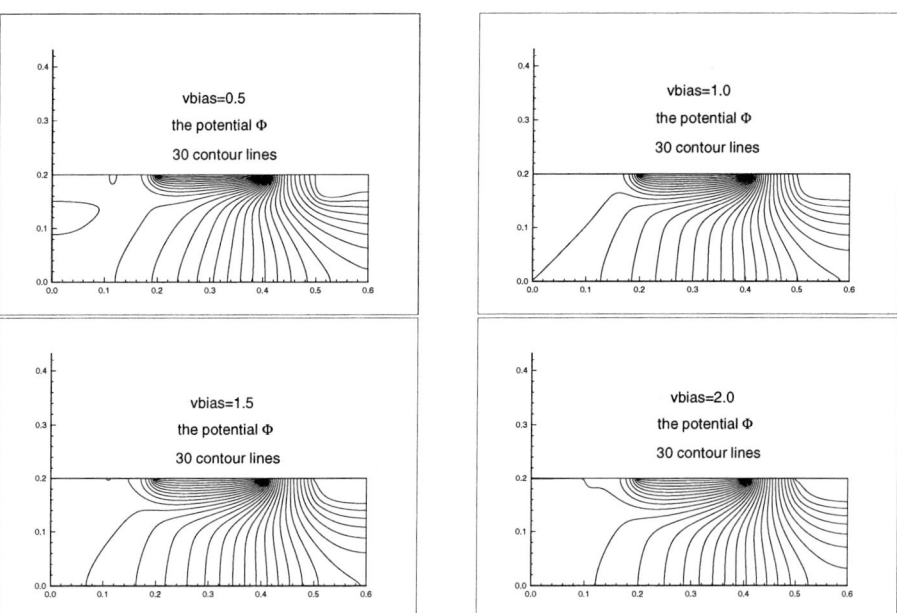

FIG. 9. Two-dimensional MESFET. Simulation result at $vbias = 0.5$ V, 1.0 V, 1.5 V and 2.0 V. The potential ϕ.

We can see from these figures that, for larger values of $vbias$, only concentration is approximately spherically symmetric around the top middle point $(x, y) = (0.3, 0.2)$.

Next, we show the result of trying to use the 1D model with a spherical symmetry assumption, to approximate the 2D MESFET described in Section 1.2. We take our 1D domain from $r = 0.025$ to $r = 0.1$, measured from the top middle point at $(x, y) = (0.3, 0.2)$ downward. The boundary conditions for the concentration n, the temperature $\mathcal{T}$ and the potential ϕ are prescribed, using the values of the 2D simulations; the boundary condition for the velocity is floating (Neumann). In Fig. 10, we show the comparison, for the concentration n, of the 2D MESFET result with the 1D model assuming spherical symmetry, at $vbias = 0.5$ V, 1.0 V, 1.5 V, and 2.0 V. We can clearly see a qualitatively correct agreement. This is very promising since it means that other quantities (such as $\mathcal{T}$ and ϕ) which are not spherically symmetric have minimal effect on the concentration through the nonlinear coupling of the equations.

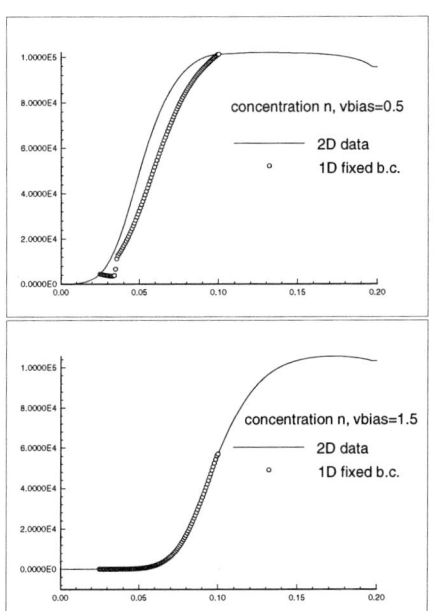
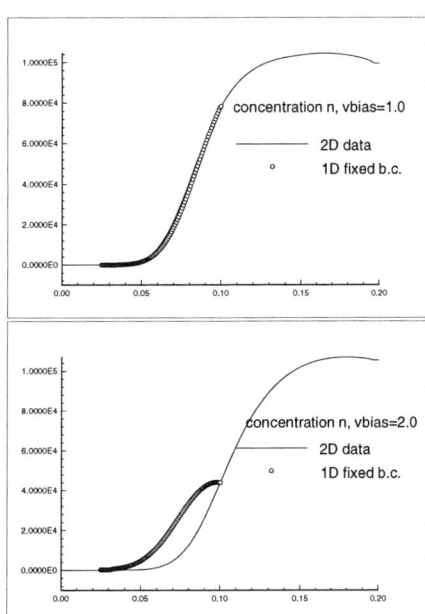

FIG. 10. The 1D model with spherical symmetry assumption, in comparison with the 2D MESFET results, at $vbias = 0.5$ V, 1.0 V, 1.5 V and 2.0 V. The concentration n.

Next, we show the same comparison for the temperature $\mathcal{T}$ in Fig. 11. We can see that now the 1D model is at much greater variance with the 2D results, manifesting the fact that $\mathcal{T}$ is not spherically symmetric. Pictures for v and ϕ show similar discrepancies. If n is the only quantity of interest, then the 1D model can be used, saving substantial computing time in the simulation. Otherwise, a better model (perhaps a hybrid one with the non-symmetric components computed by the 2D model and symmetric quantities computed by the 1D model) might be useful.

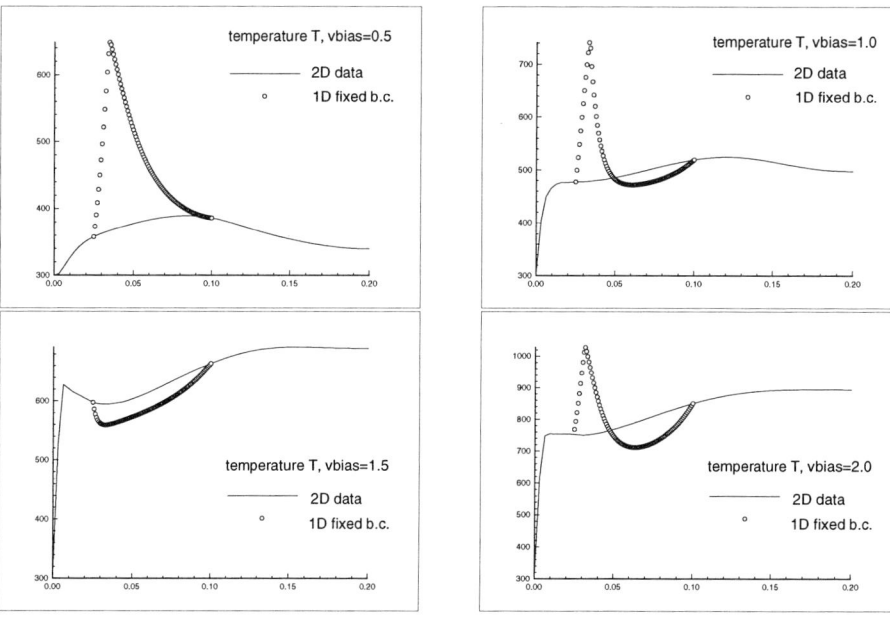

FIG. 11. The 1D model with spherical symmetry assumption, in comparison with the 2D MESFET results, at $vbias = 0.5$ V, 1.0 V, 1.5 V and 2.0 V. The temperature $\mathcal{T}$.

Acknowledgements

We would like to thank Professor Umberto Ravaioli for his help in formulating the coupling terms for the two-valley hydrodynamic model used in the Gunn oscillator. The first author is supported by the Office of Naval Research under grant N00014-91-J-1384, the National Science Foundation under grant DMS-9207080, and by an Alfred P. Sloan Foundation Fellowship. The second author is supported by the National Science Foundation under grant DMS-9424464. The third author is supported by the National Science Foundation under grant ECS-9214488 and the Army Research Office under grant DAAH04-94-G-0205. Computation is supported by the Pittsburgh Supercomputer Center.

Bibliography

1. Bløtekjær, K. (1970). Transport equations for electrons in two–valley semiconductors. *IEEE Trans. Electron Devices*, **17**, 38–47.
2. Chen, G.-Q. (1988). Convergence of [the] Lax-Friedrichs scheme for isentropic gas dynamics, III. *Acta Math. Scientia*, **8**, 243–276. (Chinese version: 1986, **6**, 75–120).
3. Chen, G.-Q. (1990). The compensated compactness method and the system of isentropic gas dynamics. *MSRI Preprint 00527-91*. Berkeley.
4. Chen, G.-Q. (1997). Remarks on spherically symmetric solutions to the compressible Euler equations. *Proc. Royal Soc. Edinburgh*, **127A**, 243–259.
5. Chen, G.-Q. and Glimm, J. (1996). Global solutions to the compressible Euler equations with geometrical structure. *Commun. Math. Phys.*, **180**, 153–193.
6. Chen, G.-Q. and Glimm, J. (1996). Global solutions to the cylindrically symmetric rotating motion of isentropic gases. *Z. Angew. Math. Phys.*, **47**, 353–372.
7. Chen, G.-Q., Jerome, J. W. and Zhang, B. (1997). Particle hydrodynamic moment models in biology and microelectronics: singular relaxation limits. *Nonlinear Anal.*, **30**, 233–244.
8. Chen, G.-Q. and Wang, D. (1996). Convergence of shock capturing schemes for the compressible Euler-Poisson equations. *Commun. Math. Phys.*, **179**, 333–364.
9. Curow, M. and Hintz, A. (1987). Numerical simulation of nonstationary electron transport in Gunn devices in an harmonic mode oscillator circuit. *IEEE Trans. Electr. Dev.*, **34**, 1983–1994.
10. Dafermos, C. (1972). Polygonal approximations of solutions of the initial-value problem for a conservation law. *J. Math. Anal. Appl.*, **38**, 33–41.
11. Ding, X., Chen, G.-Q., and Luo, P. (1987/88). Convergence of the Lax-Friedrichs scheme for isentropic gas dynamics (I)–(II). *Acta Math. Scientia*, **7**, 467–480; **8**, 61–94. (Chinese version: 1985, **5**, 415–432, 433–472).
12. DiPerna, R. (1983). Convergence of the viscosity method for isentropic gas dynamics. *Commun. Math. Phys.*, **91**, 1–30.
13. Embid, P., Goodman, J., and Majda, A. (1984). Multiple steady states for 1-D transonic flow. *SIAM J. Sci. Stat. Comp.*, **5**, 21–41.
14. Fang, W. and Ito, K. (1997). Weak solutions to a one-dimensional hydrodynamic model of two carrier types for semiconductors. *Nonlinear Anal.*, **28**, 947–963.
15. Glaz, H. and Liu, T.-P. (1984). The asymptotic analysis of wave interactions and numerical calculation of transonic nozzle flow. *Adv. Appl. Math.*, **5**, 111–146.

16. Glimm, J., Marshall, G. and Plohr, B. (1984). A generalized Riemann problem for quasi-one-dimensional gas flow. *Adv. Appl. Math.*, **5**, 1–30.
17. Jerome, J. W. (1996). *Analysis of Charge Transport: A Mathematical Study of Semiconductor Devices.* Springer.
18. Jerome, J. W. and Shu, C.-W. (1994). Energy models for one-carrier transport in semiconductor devices. In: *IMA Volumes in Mathematics and Its Applications*, vol. 59 (ed. W. Coughran, J. Cole, P. Lloyd, and J. White), pp. 185–207. Springer.
19. Jerome, J. W. and Shu, C.-W. (1995). Transport effects & characteristic modes in the modeling & simulation of submicron devices. *IEEE Trans. on Computer–Aided Design*, **14**, 917–923.
20. Jerome, J. W. and Shu, C.-W. (1995). The response of the hydrodynamic model to heat conduction, mobility, and relaxation expressions. *VLSI Design*, **3**, 131–143.
21. Lax, P. D. (1973). *Hyperbolic Systems of Conservation Laws and the Mathematical Theory of Shock Waves.* C.B.M.S. 11. SIAM.
22. Lee, C. H. and Ravaioli, U. (1991). Transient Monte-Carlo simulation of heterojunction microwave oscillators. In: *Computational Electronics* (ed. K. Hess, J. P. Leburton, and U. Ravaioli), pp. 169–172. Kluwer Academic Publishers.
23. Liu, T.-P. (1979). Quasilinear hyperbolic systems. *Comm. Math. Phys.*, **68**, 141–172.
24. Liu, T.-P. (1987). Nonlinear resonance for quasilinear hyperbolic equations. *J. Math. Phys.*, **28**, 2593–2602.
25. Makino, T. and Takeno, S. (1994). Initial boundary value problem for the spherically symmetric motion of isentropic gas. *Japan J. Indust. Appl. Math.*, **11**, 171–183.
26. Natalini, R. (1996). The bipolar hydrodynamic model for semiconductors and the drift-diffusion equations. *J. Math. Anal. Appl.*, **198**, 262–281.
27. Selberherr, S. (1984). *Analysis & Simulation of Semiconductor Devices.* Springer-Verlag, Wien–New York.
28. Shu, C.-W. and Osher, S. J. (1989). Efficient implementation of essentially non-oscillatory shock capturing schemes, II. *J. Comp. Phys.*, **83**, 32–78.
29. Tartar, L. (1979). *Compensated compactness and applications to partial differential equations*, Research Notes in Mathematics, Nonlinear Analysis and Mechanics, vol. 4 (ed. R. J. Knops). Pitman Press.

7
Approximation Issues for Applications in Optimal Control and Parameter Estimation

H. T. Banks
North Carolina State University

and

R. H. Fabiano
University of North Carolina at Greensboro

Abstract

We consider computational aspects of using semidiscrete approximation schemes to solve problems with infinite-dimensional dynamics. We survey theory and convergence results for the simulation or forward problem, the feedback control problem, and the parameter estimation problem, all for the case in which the underlying system dynamics is governed by a partial differential equation. In particular we investigate the critical sufficient conditions required for convergence of semidiscrete approximations of these problems. These sufficient conditions require that the approximation scheme demonstrate system convergence, adjoint system convergence, and uniform preservation of system stability. By considering in detail several specific examples, we illustrate the difficulties which may arise when these sufficient conditions are not satisfied.

1 Introduction

In this chapter we consider computational issues which might arise when one attempts to "solve" a problem involving infinite-dimensional dynamics (i.e., partial differential equations, delay equations, etc.) by solving instead an approximate problem involving finite-dimensional dynamics. The main point we wish to make is that a carefully constructed approximation scheme which performs well on one type of problem (a simulation problem, for example) may perform poorly on another type of problem (an inverse or optimization problem, for example), even when the underlying dynamics is the same for both problems.

We restrict our consideration to problems in which the underlying dynamics is governed by the following Cauchy problem evolving in an infinite-dimensional Hilbert space H:

$$\begin{aligned} \dot{z}(t) &= A(q)z(t) + Bu(t) \\ z(0) &= z_0. \end{aligned} \quad (1.1)$$

Here $A(q)$ is a parameter-dependent linear operator (typically unbounded), and $Bu(t)$ represents a control input or an external force. We consider solution of three different types of problems associated with the dynamics modeled by (1.1). These can be described loosely as follows:

- *Simulation* — given z_0, q, B and $u(t)$, determine the unique solution $z(t)$.
- *Parameter Estimation* — given z_0, B, $u(t)$ and data consisting of (perhaps partial) measurements of $z(t)$ at certain times t_i, estimate q.
- *Optimal Control* — given z_0, q, B and a cost functional J, determine $u(t)$ such that J is minimized.

Considering for now only the simulation problem, we know that one way to "solve" such a problem is to introduce a semidiscrete approximation scheme, whereby instead of (1.1) one solves the finite-dimensional system

$$\begin{aligned}\dot{z}^N(t) &= A^N(q)z^N(t) + B^N u(t) \\ z^N(0) &= z_0^N.\end{aligned} \quad (1.2)$$

The idea is that since (1.2) is theoretically a solved problem, we are finished "solving" the simulation problem once we have demonstrated appropriate convergence of $z^N(t)$ to $z(t)$. In this context, when we say that (1.2) is a solved problem we mean that there are known, reasonably well-behaved computational algorithms with which one can solve (1.2) on a computer (actually one considers a matrix representation of (1.2) to be solved computationally). This usually turns out to be a highly satisfactory method for solving the simulation problem.

It seems reasonable, then, to attempt the same methodology for solving the parameter estimation and/or optimal control problems, especially since the finite-dimensional versions of the parameter estimation/optimal control problems are (theoretically) solved. This is, in fact, a common approach to solving infinite-dimensional optimization problems and is sometimes described heuristically as "first discretize, then optimize." It has been observed, however, and this is precisely the main theme of this chapter, that caution must be exercised in choosing discretization schemes and optimization algorithms which are "compatible". Indeed it turns out that an excellent discretization scheme (excellent for simulation, that is) and an excellent optimization algorithm may perform poorly when used together.

In the remainder of the chapter we will expand on these ideas. More detailed definitions and statements of convergence theorems will be given so that we may focus more precisely on the theoretical reasons for the behavior which has been described. In Section 2 we consider the simulation problem and recall some standard definitions and convergence results. In Section 3 we survey theoretical results for the linear-quadratic regulator problem (an optimal control problem). We will focus on recent results for two specific partial differential equations, one a model for a delay system and the other a model for an elastic structure with boundary damping. Finally in Section 4 we pursue similar ideas for the parameter estimation problem.

2 The Simulation Problem

In this section we recall some standard definitions and results for the simulation problem. Specifically, consider a linear Cauchy problem

$$\dot{z}(t) = Az(t) + f(t) \qquad (2.1)$$
$$z(0) = z_0$$

on an infinite-dimensional Hilbert space H. We assume that $A:\mathrm{dom}A \subset H \to H$ is the infinitesimal generator of a C_0-semigroup $T(t)$ on H. This assumption guarantees that (2.1) is well-posed (we take as a standing assumption throughout the paper that the underlying dynamics or "forward" problem is well-posed). The simulation problem consists of finding $z(t)$ when z_0 and $f(t)$ are known. Proceeding as described in the Introduction, we introduce a semidiscrete approximation scheme for (2.1), which will consist of a sequence of finite-dimensional subspaces $H^N \subset H$, $N = 1, 2, \ldots$, the corresponding orthogonal projections $P^N : H \to H^N$, and operators $A^N : H^N \to H^N$. This defines a finite dimensional analog of (2.1) given by

$$\dot{z}^N(t) = A^N z^N(t) + P^N f(t) \qquad (2.2)$$
$$z(0) = P^N z_0.$$

The operator A^N generates the semigroup $T^N(t) = e^{tA^N}$ on H^N. The system (2.2) is equivalent to a first-order matrix differential equation which can be solved computationally, and from this the solution $z^N(t)$ of (2.2) can be recovered. Many standard approximation methods can be formulated in this semigroup-theoretic framework, including Galerkin, finite element, spectral, and finite difference methods. In order for this approach to be a useful method for solving the simulation problem, we need a result guaranteeing convergence of solutions of (2.2) to solutions of (2.1). The following hypotheses are sufficient.

(H1) For each $z \in H$, $P^N z \to z$ in H (that is, P^N converges strongly to P).

(H2) For each $z \in H$, $T^N(t)P^N z \to T(t)z$ in H, uniformly on compact t-intervals.

Recall that the mild solution of (2.1) can be written (using the variation of constants formula) as

$$z(t) = T(t)z_0 + \int_0^t T(t-s)f(s)\,ds.$$

Similarly, the variation of constants solution of (2.2) is given by

$$z^N(t) = T^N(t)P^N z_0 + \int_0^t T^N(t-s)P^N f(s)\,ds.$$

Thus, we see that if (H1) and (H2) hold, then $z^N(t) \to z(t)$ in H, uniformly on compact t-intervals. In many applications, H is a function space (or cross product of function spaces), H^N is the span of a finite number of basis functions, and N is related to the number of basis functions and/or the "mesh size". Typically, hypothesis (H1) follows from the approximation properties of the basis functions. A useful tool for verifying (H2) is the Trotter–Kato theorem, a semigroup-theoretic version of the well-known dictum [32] "stability plus consistency yields convergence". For various versions and applications of the Trotter–Kato theorem, see [2, 7, 9, 26, 34]. In summary, in order to solve the simulation problem (2.1) via the semidiscrete approximation (2.2), the scheme must have the properties that the finite-dimensional spaces H^N approximate well the space H (hypothesis (H1)), and that the finite-dimensional operators A^N are a stable and consistent approximation of the operator A (which yields hypothesis (H2)).

3 The Optimal Control Problem

In this section we consider computational issues for optimal control problems, and in particular we focus on the linear-quadratic regulator (LQR) problem. To set up the problem, let H, U, and W be Hilbert spaces representing respectively the state space, the control space, and the output space. Consider a cost functional

$$J(z_0, u) = \int_0^\infty [\langle Cz(t), Cz(t)\rangle_W + \langle Ru(t), u(t)\rangle_U]\, dt \qquad (3.1)$$

and the state equation

$$\begin{aligned} \dot{z}(t) &= Az(t) + Bu(t) \\ z(0) &= z_0. \end{aligned} \qquad (3.2)$$

As in the simulation problem, we assume that A: dom$A \subset H \to H$ is the infinitesimal generator of a C_0-semigroup $T(t)$ on H. We also assume that the weight operators in the cost functional satisfy $C \in \mathcal{L}(H, W)$ and $R \in \mathcal{L}(U)$, with $R > 0$. (We use the standard notation $\mathcal{L}(H_1, H_2)$ to denote the space of bounded linear operators from H_1 to H_2, and $\mathcal{L}(H_1)$ if $H_1 = H_2$). For simplicity of presentation we assume that the control operator is bounded, that is, $B \in \mathcal{L}(U, H)$, although it should be noted that many of the results summarized below are available for the case in which B and/or C is unbounded. (For a recent summary of some of these results, see Chapter 7 of [11].) With these definitions, we can state the LQR problem of interest, which is

$$\min_{u \in L_2(0,\infty;U)} J(z_0, u) \qquad (3.3)$$

subject to dynamics governed by (3.2). Among many references considering this problem, we cite [16] and [22].

As we did with the simulation problem, we will first recall some conditions which guarantee that the LQR problem is "well-posed" (here this means that there exist a unique control given in feedback form). Then we will introduce a semidiscrete approximation of the LQR problem, and consider necessary conditions for convergence. To proceed, let us recall the following definitions relevant to the LQR problem.

▷ The C_0-semigroup $T(t)$ is *exponentially stable* if there exists $M \geq 1$ and $\omega > 0$ such that $\|T(t)\| \leq Me^{-\omega t}$ for all $t \geq 0$.
▷ The pair (A, B) is *stabilizable* if there exists $K \in \mathcal{L}(H, U)$ such that $A - BK$ is the infinitesimal generator of an exponentially stable semigroup.
▷ The pair (A, C) is *detectable* if there exists $L \in \mathcal{L}(W, H)$ such that $A - LC$ is the infinitesimal generator of an exponentially stable semigroup.
▷ The control $u \in L_2(0, \infty; U)$ is *admissible for the initial condition* $z_0 \in H$ if $J(z_0, u) < \infty$.
▷ The system (3.1)–(3.2) is *optimizable* if there exists an admissible control u for every $z_0 \in H$.

Note that if (A, B) is stabilizable then the system (3.1)–(3.2) is optimizible, and if A generates an exponentially stable semigroup, then (A, B) is stabilizable and (A, C) is detectable. Stabilizability and detectability hypotheses are important sufficient conditions for well-posedness of the LQR problem, as the following well-known result indicates.

Theorem 3.1 *For the system (3.1)–(3.2), there exists a nonnegative self-adjoint solution* $\Pi \in \mathcal{L}(H)$ *of the algebraic Riccati equation (ARE)*

$$A^*\Pi + \Pi A - \Pi B R^{-1} B^* \Pi + C^* C = 0 \tag{3.4}$$

if and only if (3.1)–(3.2) is optimizable. In this case, if Π *is the minimal nonnegative self-adjoint solution of the ARE (3.4), then the unique optimal control for the LQR problem is given in feedback form by*

$$\widetilde{u}(t) = -R^{-1} B^* \Pi \widetilde{z}(t), \tag{3.5}$$

where the optimal trajectory $\widetilde{z}(t)$ *is the solution of the closed loop system*

$$\begin{aligned} \dot{z}(t) &= (A - BR^{-1}B^*\Pi)z(t) \\ z(0) &= z_0. \end{aligned} \tag{3.6}$$

That is, $\widetilde{z}(t) = S(t)z_0$, *where* $S(t)$ *is the closed loop semigroup generated by* $A - BR^{-1}B^*\Pi$. *If in addition* (A, C) *is detectable, then there exists a unique nonnegative self-adjoint solution of the ARE (3.4), and* $S(t)$ *is exponentially stable.*

An alternative useful hypothesis is the following.

(H3) For each $z_0 \in H$, there exists an admissible control, and any admissible control drives the state $z(t)$ to zero as $t \to \infty$ (that is, $\lim_{t\to\infty} \|z(t)\|_H = 0$).

If (H3) holds, then there exists a unique nonnegative self-adjoint solution of the ARE (3.4). If $C^*C > 0$, then (H3) holds and $S(t)$ is exponentially stable.

From Theorem 3.1 we see that the optimal control is given in feedback form by $u = Kx$, where $K = -R^{-1}B^*\Pi$ is sometimes referred to as the feedback gain operator. In solving the LQR problem, we may be interested in the closed loop solution $\tilde{z}(t)$, the closed loop semigroup $S(t)$, the control $u(t)$, the solution Π of the ARE, or perhaps just the gain K. To find any of these computationally, we need a semidiscrete approximation scheme and appropriate convergence results.

Therefore, consider a semidiscrete approximation scheme consisting of a sequence of finite-dimensional subspaces $H^N \subset H$, $N = 1, 2, \ldots$, the corresponding orthogonal projections $P^N : H \to H^N$, and operators $A^N : H^N \to H^N$. Also let $B^N \in \mathcal{L}(U, H^N)$ and $C^N \in \mathcal{L}(H^N, W)$. This allows one to define an approximate cost functional

$$J^N(P^N z_0, u) = \int_0^\infty [\langle C^N z(t), C^N z(t)\rangle_W + \langle Ru(t), u(t)\rangle_U]\, dt \quad (3.7)$$

and finite-dimensional state equation

$$\begin{aligned} \dot{z}^N(t) &= A^N z^N(t) + B^N u(t) \\ z^N(0) &= P^N z_0. \end{aligned} \quad (3.8)$$

The finite-dimensional LQR problem is

$$\min_{u \in L_2(0,\infty;U)} J^N(P^N z_0, u) \quad (3.9)$$

subject to dynamics governed by (3.8). As in the infinite dimensional case, there are various hypotheses which guarantee well-posedness for the finite-dimensional LQR problem. Whichever hypotheses are used, however, should be satisfied for all N. The following result is typical.

Theorem 3.2 *Assume that (H3) holds for each N. Then there exists a unique nonnegative self-adjoint solution Π^N of the algebraic Riccati equation (ARE)*

$$A^{N*}\Pi^N + \Pi^N A^N - \Pi^N B^N R^{-1} B^{N*}\Pi^N + C^{N*}C^N = 0. \quad (3.10)$$

Moreover, the unique optimal control for the finite-dimensional LQR problem is given in feedback form by

$$\tilde{u}^N(t) = -R^{-1}B^{N*}\Pi^N \tilde{z}^N(t), \quad (3.11)$$

where the optimal trajectory $\tilde{z}^N(t)$ is the solution of the closed loop system

$$\begin{aligned} \dot{z}^N(t) &= (A^N - B^N R^{-1} B^{N*}\Pi^N) z^N(t) \\ z^N(0) &= P^N z_0. \end{aligned} \quad (3.12)$$

That is, $\tilde{z}^N(t) = S^N(t)P^N z_0$, where $S^((t) = e^{(A^N - B^N R^{-1} B^{N*} \Pi^N)t}$ is the closed loop semigroup generated by $A^N - B^N R^{-1} B^{N*} \Pi^N$.

We are interested in conditions which will be sufficient to guarantee appropriate convergence for these problems. In particular, we are interested in ensuring that $\Pi^N \to \Pi$ in an appropriate sense as $N \to \infty$. Based on our discussions so far, it seems reasonable to expect to need (H1)–(H3), as well as some sort of convergence of B^N to B and C^N to C. These are stated as:

(H4) (a) for each $v \in U$, $B^N v \to Bv$; (b) For each $z \in H$, $B^{N*} P^N z \to B^* z$.

(H5) For each $z \in H$, $C^N P^N z \to Cz$ and for each $w \in W$, $C^{N*} w \to C^* w$.

If (H1)–(H5) were sufficient for the convergence we desire, then there would be little point to this chapter. Indeed, these hypotheses are not sufficient, and most theoretical convergence results in the literature require what amounts to two additional hypotheses. One is an "adjoint convergence" condition, and the other is a "preservation of stability under approximation" condition. The adjoint condition can be given as follows.

(H6) For each $z \in H$, $T^N(t)^* P^N z \to T(t)^* z$ in H, uniformly on compact t-intervals.

Here $T(t)^*$ is the Hilbert space adjoint of the operator $T(t)$, and it is well known [34] that A^* is the infinitesimal generator of $T(t)^*$. Condition (H6) is not unexpected once we note that the control $u(t)$ (resp. $u^N(t)$) is defined in terms of the operator Π (resp. Π^N), which is in turn a solution of an algebraic Riccati equation involving A^* (resp. A^{N*}).

It is important to consider (H6) when designing approximation schemes for the LQR problem. In particular, (H6) is independent of (H2), although it should be noted that (H2) implies weak operator convergence of $T^N(t)^* P^N$ to $T(t)^*$ ((H6) is strong operator convergence). We shall see below an example of an approximation scheme which satisfies (H2) but not (H6), and which performs well for the simulation problem but poorly for the control problem. If one were not familiar with the results surveyed here, this type of behavior might be rather surprising.

The "preservation of stability under approximation" condition appears in various forms in the literature. The following result is typical and was given in [8].

Theorem 3.3 *Suppose (H1), (H2), (H4), (H5), (H6) hold, and (H3) holds for each N. (Note that this guarantees that the finite-dimensional LQR problem is well-posed for each N). Let Π^N denote the unique nonnegative self-adjoint solution of the ARE (3.10). Further assume that the ARE (3.4) has a unique nonnegative self-adjoint solution Π. Let $S(t)$ and $S^N(t)$ denote the closed loop* semigroups generated by $A - BR^{-1}B^*\Pi$ and $A^N - B^N R^{-1} B^{N*} \Pi^N$, respectively. IF there are positive constants M_1, M_2 and ω (independent of N) satisfying

$$\|S^N(t)\|_{H^N} \leq M_1 e^{-\omega t} \quad \text{for } t \geq 0 \tag{3.13}$$

and

$$\|\Pi^N\|_{H^N} \leq M_2, \tag{3.14}$$

for all N, then

$$\Pi^N P^N z \to \Pi z \quad \text{for every } z \in H, \tag{3.15}$$

$$S^N(t) P^N z \to S(t) z \quad \text{for every } z \in H, \tag{3.16}$$

where the convergence is uniform in t on bounded intervals, and

$$\|S(t)\| \leq M_1 e^{-\omega t} \quad \text{for } t \geq 0. \tag{3.17}$$

In the statement of this theorem, one interprets (3.13) as a uniform stability condition. Unfortunately, for many approximation schemes it is not easy to verify (3.13) and (3.14). Assuming that the control space U is finite-dimensional, we can sometimes instead use a related result, which makes use of the following definitions.

▷ Given an approximation scheme as described above, we say that (A^N, B^N) is *uniformly stabilizable* if there exists $K^N \in \mathcal{L}(H^N, U)$, $M \geq 1$ and $\omega > 0$ such that $\sup \|K^N\| < \infty$ and $\|e^{(A^N - B^N K^N)t} P^N\| \leq M e^{-\omega t}$ for all $t \geq 0$ and N sufficiently large.

▷ Given an approximation scheme as described above, we say that (A^N, C^N) is *uniformly detectable* if there exists $L^N \in \mathcal{L}(W, H^N)$, $M \geq 1$ and $\omega > 0$ such that $\sup \|L^N\| < \infty$ and $\|e^{(A^N - L^N C^N)t} P^N\| \leq M e^{-\omega t}$ for all $t \geq 0$ and N sufficiently large.

Theorem 3.4 (see [25]) *Suppose (H1), (H2), (H4), (H5) and (H6) hold, and that (A^N, B^N) is uniformly stabilizable and (A^N, C^N) is uniformly detectable. Then for each N the finite-dimensional ARE (3.10) has a unique nonnegative solution Π^N and there exists $M_1 \geq 1$, $\omega > 0$ such that (3.13) holds. Moreover, (A, B) is stabilizable. If in addition (A, C) is detectable, then the ARE (3.4) has a unique nonnegative self-adjoint solution Π, and (3.15) holds.*

It is possible to obtain (3.15) under weaker versions of the "preservation of stability under approximation" condition. For example, in [30] the authors define *uniform output stability* and *uniform input–output stability* for approximation schemes and show that these conditions (which are weaker than uniform stabilizability and detectability) together with (H1)–(H6) imply (3.15). Alternatively,

Approximation Issues for Applications

in some applications it happens that the open-loop semigroup $T(t)$ is exponentially stable, and if the approximation scheme preserves this property (as characterized in the following definition) a convergence result can be obtained for the LQR problem.

▷ Assume that $T(t)$ is exponentially stable. Given an approximation scheme as described above for the LQR problem, we say that $T^N(t) = e^{A^N t}$ is *uniformly exponentially stable* if there exists $M \geq 1$ and $\omega > 0$ such that $\sup \|T^N(t)\| \leq M e^{-\omega t}$ for all $t \geq 0$ and N sufficiently large.

Assuming that $T(t)$ is exponentially stable, $T^N(t)$ is uniformly exponentially stable, and (H1)-(H6) hold, it can be inferred from the results in [23] and [8] that (3.15) and (3.16) hold.

We turn next to a couple of examples in which the "adjoint convergence" and "preservation of stability under approximation" conditions will be explored for certain specific approximation schemes. These examples illustrate in very specific ways why the theoretical concepts and issues raised in the preceding discussions are indeed of computational importance. Difficulties that arise when certain of the sufficient conditions are not met will be rather apparent.

Example 1 Consider a cost functional

$$J(u) = \int_0^\infty \left[y(t,0)^2 + u(t)^2 \right] dt$$

and state equation

$$\frac{\partial y}{\partial t}(t,s) = \frac{\partial y}{\partial s}(t,s), \quad s \in (-1,0), \quad t > 0,$$

$$\frac{\partial y}{\partial t}(t,0) = y(t,0) + y(t,-1) + u(t),$$

$$y(0,s) = \Phi(s), \quad -1 \leq s \leq 0.$$

This system arises in the study of delay equations (in fact, by setting $x(t+s) = y(t,s)$ the system is equivalent to the delay equation $\dot{x}(t) = x(t) + x(t-1) + u(t)$, $t > 0$). In order to formulate an abstract LQR problem such as we have considered, define the state space $H = \mathbb{R}^1 \times L_2(-1,0)$, the control space $U = \mathbb{R}^1$, and the state $z(t) = (y(t,0), y(t,\cdot))$. Then the above system can be written as

$$\dot{z}(t) = Az(t) + Bu(t),$$
$$z(0) = z_0 = (\Phi(0), \Phi),$$

where $B \in \mathcal{L}(U, H)$ is defined by $Bu = (u, 0)$, and A is defined on the domain

$$\text{dom } A = \{(\eta, \phi) \in H \mid \phi \in H^1(-1, 0), \ \phi(0) = \eta\}$$

by $A(\phi(0),\phi) = (\phi(0) + \phi(-1), \frac{d}{ds}\phi)$. If we take $R = 1$ and $C \in \mathcal{L}(H)$ is given by $C(\eta,\phi) = (\eta,0)$, then the cost functional J can be written as

$$J(u) = \int_0^\infty \{\langle Cz(t), Cz(t)\rangle_H + R|u(t)|^2\}\, dt.$$

We can consider the simulation problem (2.1) and the LQR problem (3.3), both of which have received much attention in the literature. It is known (see [2] and [16]) that A is the infinitesimal generator of a C_0-semigroup $T(t)$ on H, so that the simulation problem is well-posed. Also (see [16], [5] and [23]) the LQR problem is well-posed and the optimal control is given in feedback form by (3.5). Because of the product space structure of H, one can write the operator solution Π of the ARE (3.4) as

$$\Pi = \begin{pmatrix} \Pi_{00} & \Pi_{01} \\ \Pi_{10} & \Pi_{11} \end{pmatrix},$$

where Π_{00} is a scalar (i.e., in $\mathcal{L}(\mathbb{R}^1)$), $\Pi_{11} \in \mathcal{L}(L_2(-1,0))$, $\Pi_{10} \in \mathcal{L}(\mathbb{R}^1, L_2(-1.0))$ and $\Pi_{01} \in \mathcal{L}(L_2(-1,0), \mathbb{R}^1)$. It follows from the structure of B, C and R that the optimal feedback control has the form

$$\widetilde{u}(t) = -\left(\Pi_{00}\widetilde{y}(t,0) + \int_{-1}^0 \Pi_{10}(s)\widetilde{y}(t,s)\, ds\right),$$

where $\widetilde{z}(t) = (\widetilde{y}(t,0), \widetilde{y}(t,\cdot))$ is the solution of the closed loop system (3.6). In particular, the feedback gain kernel $\Pi_{10}(s)$ is a function in $L_2(-1,0)$.

Let us now consider approximation for the simulation and LQR problems. Recall that an approximation scheme for the simulation problem consists of constructing a sequence of finite-dimensional spaces $H^N \subset H$ and operators $A^N : H^N \to H^N$. The orthogonal projections $P^N : H \to H^N$ are defined from the H^N. The sequence $\{A^N, H^N, P^N\}$ thus defines an approximation scheme for the simulation problem, and it is natural to use this scheme for the LQR problem by simply defining $B^N = P^N B$ and $C^N = P^N C$.

For the particular system under consideration, constructing H^N amounts to discretizing the function space $L_2(-1,0)$, and defining A^N amounts to discretizing the differential operator d/ds. The paper [2] is one of the earliest to consider approximation for this problem within the context of semigroup-theoretic convergence results. In it Banks and Burns considered the so-called averaging approximation scheme (referred to here as the AVE scheme), using piecewise constant functions to discretize $L_2(-1,0)$ and finite differencing to discretize the operator d/ds. Specifically, let $t_j^N = -j/N$, $j = 0, 1, \ldots, N$ be a partition of the interval $[-1, 0]$, and let $\chi_j^N = \chi_{[t_j^N, t_{j-1}^N]}$ be the usual characteristic function for the interval $[t_j^N, t_{j-1}^N]$. Define the basis elements $e_{AVE}^{N,0} = (1, 0)$ and $e_{AVE}^{N,j} = (0, \chi_j^N)$, $j = 1, 2, \ldots, N$, and set

$$H_{AVE}^N = \mathrm{span}\{e_{AVE}^{N,0}, e_{AVE}^{N,1}, \ldots, e_{AVE}^{N,N}\}.$$

Then $H_{AVE}^N \subset H$, and $A_{AVE}^N : H_{AVE}^N \to H_{AVE}^N$ is defined by

$$A_{AVE}^N \sum_{j=0}^N \alpha_j e_{AVE}^{N,j} = (\alpha_0 + \alpha_N) e_{AVE}^{N,0} + \sum_{j=1}^N N(\alpha_{j-1} - \alpha_j) e_{AVE}^{N,j}.$$

Banks and Burns verified that (H1) and (H2) hold for the AVE scheme, and used it to solve some simulation and control problems (but not the LQR problem). They also obtained a convergence rate for the simulation problem, establishing that the scheme is order $1/N$.

In an effort to improve on the convergence rate for the simulation problem, Banks and Kappel [7] considered a Galerkin-type approximation scheme (denoted as the SPL scheme) based on using piecewise spline functions to discretize $L_2(-1, 0)$. Specifically, define the piecewise linear functions

$$b_0^N(s) = \begin{cases} \frac{N}{r}(s - t_1^N), & \text{if } t_1^N \leq s \leq 0, \\ 0, & \text{elsewhere,} \end{cases}$$

$$b_N^N(s) = \begin{cases} -\frac{N}{r}(s - t_{N-1}^N), & \text{if } t_N^N \leq s \leq t_{N-1}^N, \\ 0, & \text{elsewhere,} \end{cases}$$

and, for $j = 1, 2, \ldots, N-1$,

$$b_j^N(s) = \begin{cases} -\frac{N}{r}(s - t_{j-1}^N), & \text{if } t_j^N \leq s \leq t_{j-1}^N, \\ \frac{N}{r}(s - t_{j+1}^N), & \text{if } t_{j+1}^N \leq s \leq t_j^N, \\ 0, & \text{elsewhere.} \end{cases}$$

These functions are the familiar and standard 'hat' functions used in linear finite element schemes. Next define the basis elements $e_{SPL}^{N,0} = (1, b_0^N)$, $e_{SPL}^{N,j} = (0, b_j^N)$ for $j = 1, 2, \ldots, N$, and set

$$H_{SPL}^N = \text{span} \left\{ e_{SPL}^{N,0}, e_{SPL}^{N,1}, \ldots, e_{SPL}^{N,N} \right\}.$$

Then $H_{SPL}^N \subset \text{dom } A \subset H$, and $A_{SPL}^N : H_{SPL}^N \to H_{SPL}^N$ can be defined by $A_{SPL}^N = P_{SPL}^N A$, where P_{SPL}^N is the orthogonal projection of H onto H_{SPL}^N. Banks and Kappel verified that (H1) and (H2) hold for the SPL scheme and established that the scheme is order $1/N^2$. They used the scheme to solve some simulation problems, and presented computational examples verifying convergence rates for the two schemes. Not surprisingly, the SPL scheme outperformed the AVE scheme for the simulation problem.

However, when researchers began investigating the possibility of using each of these schemes to solve the LQR problem ([5, 23]), the spline-based scheme was seen to have some shortcomings. The authors of [5] applied both the AVE and SPL schemes to the LQR problem defined above, and generated approximations to the functional feedback gain Π_{10}. Recall that Π_{10} is a function defined on $(-1, 0)$, and because of the structure of each approximation scheme, the AVE

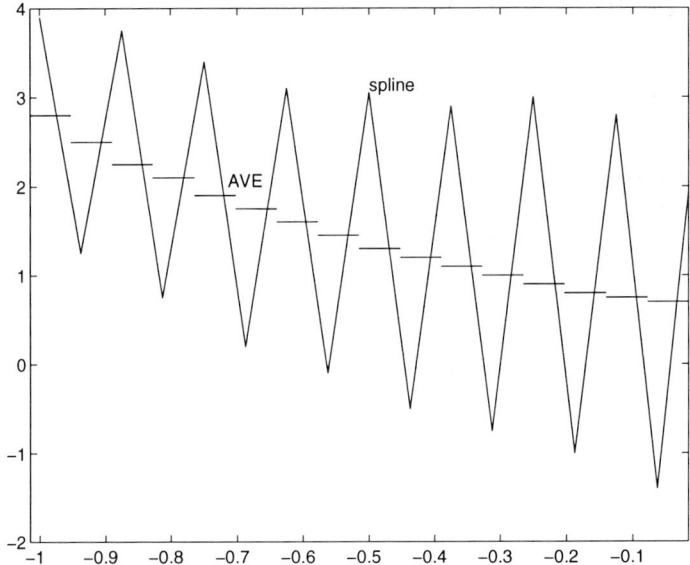

FIG. 1. Approximate feedback functional gain Π_{10}^N, $N = 16$

approximation of Π_{10} will be a piecewise constant function and the SPL approximation will be a piecewise linear function. Based on the superiority of the SPL scheme for the simulation problem, there was some expectation by the authors of [5] that the same might hold true for the LQR problem. However, Fig. 1 indicates that this is not the case. The AVE scheme provides a more useful reconstruction of Π_{10}, and Banks et al. conjectured that the SPL reconstruction only obtains some type of weak convergence for this gain. Burns et al. [12] subsequently showed that the SPL scheme does not satisfy the adjoint convergence condition (H6) (Gibson [23] showed that the AVE scheme satisfies (H6)). This lack of adjoint convergence is not too surprising when we observe that H_{SPL}^N is contained in dom A but not in dom $A^* = \{(\xi, \psi) \in H \mid \psi \in H^1(-1, 0), \psi(-1) = \xi\}$.

With this observation in mind, and in an effort to improve convergence rates for LQR approximations, Kappel and Salamon [28] defined a modified spline-based scheme (referred to as the NEW scheme) with the adjoint convergence property (H6). Specifically, using the piecewise linear functions b_j^N from above, define basis elements $e_{\text{NEW}}^{N,0} = (1, 0)$, $e_{\text{NEW}}^{N,j} = (0, b_{j-1}^N)$ for $j = 1, 2, \ldots, N+1$, and set

$$H_{\text{NEW}}^N = \text{span}\left\{e_{\text{NEW}}^{N,0}, e_{\text{NEW}}^{N,1}, \ldots, e_{\text{NEW}}^{N,N+1}\right\}.$$

Since $H_{\text{NEW}}^N \not\subset \text{dom } A$, one cannot use $P_{\text{NEW}}^N A$ to define an operator A_{NEW}^N on H_{NEW}^N (here P_{NEW}^N is the orthogonal projection of H onto H_{NEW}^N). Instead, extend

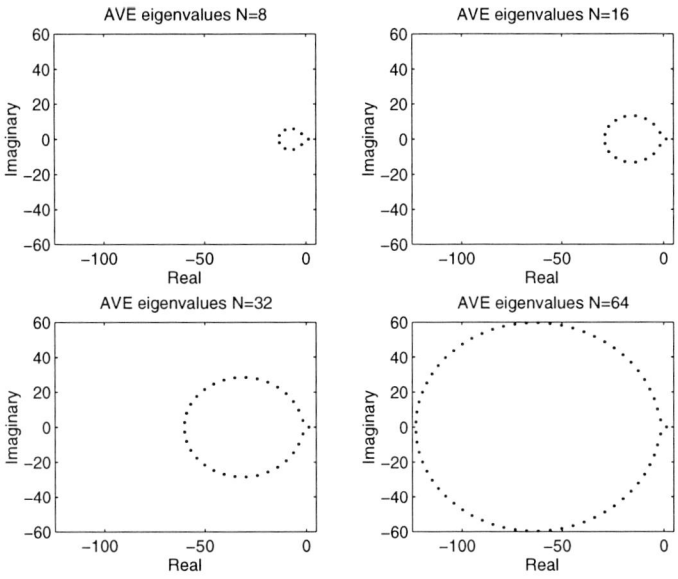

FIG. 2. Eigenvalues of A_{AVE}^N (AVE scheme)

A to the operator $\widetilde{A}$ defined on the domain $\operatorname{dom}\widetilde{A} = \mathbb{R} \times H^1(-1,0)$ by

$$\widetilde{A}(\eta, \phi) = (\eta + \phi(-1), \frac{d}{ds}\phi + \delta(s)[\eta - \phi(0)]),$$

where $\delta(s)$ denotes the Dirac delta impulse at $s = 0$. Thus $\operatorname{dom}\widetilde{A} \supset H_{\mathrm{NEW}}^N$ and $A_{\mathrm{NEW}}^N : H_{\mathrm{NEW}}^N \to H_{\mathrm{NEW}}^N$ can be defined by $A_{\mathrm{NEW}}^N = P_{\mathrm{NEW}}^N \widetilde{A}$. Kappel and Salamon [28] showed that (H1)–(H6) hold for this scheme, and observed *numerically* that for the LQR problem, the scheme converged and outperformed both the AVE and SPL schemes. They did not prove convergence *theoretically* because, as they showed, the NEW scheme does not satisfy a uniform exponential stability condition. In two later papers, Kappel and Salamon [29, 30], finally showed theoretical convergence of the NEW scheme for the LQR problem, by first showing that the NEW scheme satisfies a weaker uniform stability condition and then showing that this is enough for convergence of the gains.

Let us pursue the issue of uniform stability behavior a bit further. To visualize this property numerically, we plot the eigenvalues of A_{AVE}^N and A_{NEW}^N for several values of N in Fig. 2 and Fig. 3. We see that the AVE scheme exhibits uniform stability *numerically*, and this has been established theoretically in [36]. For the NEW scheme, it is the existence of eigenalues λ_{NEW}^N of A_{NEW}^N with the property that

$$\lim_{N \to \infty} \operatorname{Re} \lambda_{\mathrm{NEW}}^N = 0 \quad \text{and} \quad \lim_{N \to \infty} \operatorname{Im} \lambda_{\mathrm{NEW}}^N = \infty$$

that causes the lack of uniform exponential stability. The eigenvalues of the SPL scheme exhibit the same behavior. Thus even though uniform exponential

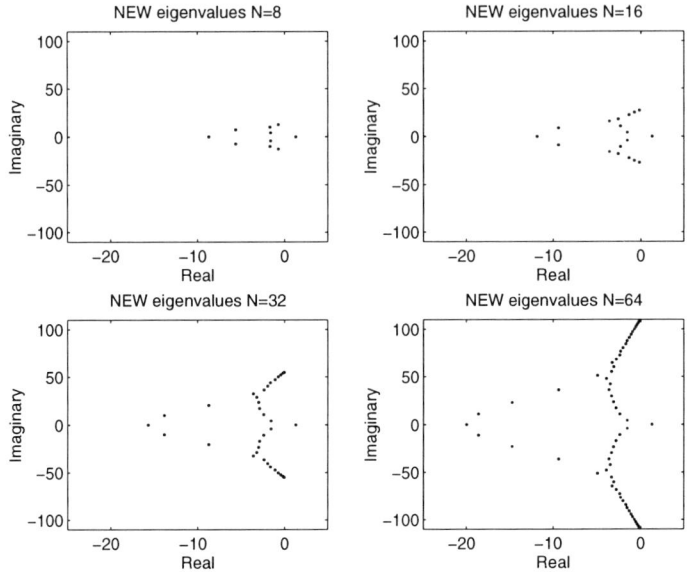

FIG. 3. Eigenvalues of A_{NEW}^N (NEW scheme)

stability was not necessary for convergence of the feedback gains in the LQR approximation problem, the lack of this property could still be considered a shortcoming of the NEW scheme. There is ongoing research activity involving the construction of approximation schemes for such delay systems, and particularly on schemes which satisfy the adjoint convergence and uniform exponential stability conditions (see [27] and references therein).

We complete our considerations of this example with a discussion of one particular recently developed method [20] for constructing Galerkin approximation schemes which satisfy a uniform stability condition. Recall that in our interpretation of the Galerkin method as used above, we constructed finite-dimensional spaces $H^N \subset \text{dom} A$ and defined $A^N : H^N \to H^N$ by $A^N = P^N A$, where P^N is the orthogonal projection of H onto H^N. A more standard interpretation of the Galerkin method is to consider a sesquilinear form $\sigma(\cdot, \cdot) : V \times V \to \mathbb{C}$ which is related to A by $\text{dom} A \subset V$ and

$$\langle Ax, y \rangle = \sigma(x, y) \quad \text{for all } x, y \in V.$$

One then constructs finite-dimensional spaces $H^N \subset V$ and defines $A^N : H^N \to H^N$ by

$$\langle A^N x, y \rangle = \sigma(x, y) \quad \text{for all } x, y \in H^N. \tag{3.18}$$

(Both the SPL and NEW schemes can be defined according to this approach to the Galerkin method). The idea in [20] is to select a more appropriate equivalent inner product $\langle \cdot, \cdot \rangle_e$ on H (two inner products are equivalent if their compatible

norms are equivalent) and corresponding sesquilinear form σ_e such that
$$\langle Ax, y \rangle_e = \sigma_e(x, y) \quad \text{for all } x, y \in V.$$
One can then use the same finite-dimensional spaces $H^N \subset V$ and define $A_e^N : H^N \to H^N$ by
$$\langle A_e^N x, y \rangle_e = \sigma_e(x, y) \quad \text{for all } x, y \in H^N. \tag{3.19}$$
For the example under consideration here, we have
$$\langle (\eta, \phi), (\xi, \psi) \rangle = \eta\xi + \int_{-1}^{0} \phi(s)\psi(s)\,ds$$
and
$$\sigma((\eta, \phi), (\xi, \psi)) = [\eta + \phi(-1)]\xi + \int_{-1}^{0} \frac{d}{ds}\phi(s)\psi(s)\,ds.$$
Using the finite-dimensional spaces H_{NEW}^N and (3.18) leads to the NEW scheme. Consider instead the inner product
$$\langle (\eta, \phi), (\xi, \psi) \rangle_e = \eta\xi + e^{-\gamma} \int_{-1}^{0} e^{-2\gamma s}\phi(s)\psi(s)\,ds,$$
where γ is the unique real eigenvalue of A, and also consider the sesquilinear form
$$\sigma_e((\eta, \phi), (\xi, \psi)) = [\eta + \phi(-1)]\xi + e^{-\gamma}\int_{-1}^{0} e^{-2\gamma s}\frac{d}{ds}\phi(s)\psi(s)\,ds + e^{-\gamma}[\eta - \phi(0)]\psi(0)$$

(see [20] for details). Then using the finite-dimensional spaces H_{NEW}^N and (3.19) defines an operator $\widehat{A}_{\text{NEW}}^N : H_{\text{NEW}}^N \to H_{\text{NEW}}^N$ which satisfies a uniform stability property (this was shown in [20]). In Fig. 4 we plot the eigenvalues of $\widehat{A}_{\text{NEW}}^N$. While the previous discussion illustrates that uniform stability is, strictly speaking, not necessary for convergence in the LQR approximation problem, it nonetheless seems to be a computationally desirable property. That is, if an operator A generates an exponentially stable semigroup $T(t)$ satisfying $\|T(t)\| \le Me^{-\omega t}$, then it would seem that an approximation scheme should consist of operators A^N with the property that $T^N(t) = e^{A^N t}$ is uniformly exponentially stable. However, the present authors are unable to make a definitive statement concerning the precise negative implications for the case that $T^N(t) = e^{A^N t}$ is *not* uniformly exponentially stable. We consider this to be an open and unresolved issue. To complete this section we will consider the issue again as it arises in a model of a weakly damped elastic system.

Example 2 Consider the following equation:
$$y_{tt}(t, \xi) = y_{\xi\xi}(t, \xi), \quad 0 < \xi < 1, \tag{3.20}$$
$$y(t, 0) = 0, \quad y_\xi(t, 1) = -\alpha y_t(t, 1), \quad \alpha > 0,$$
$$y(0, \xi) = y_0(\xi), \quad y_t(0, \xi) = v_0(\xi).$$

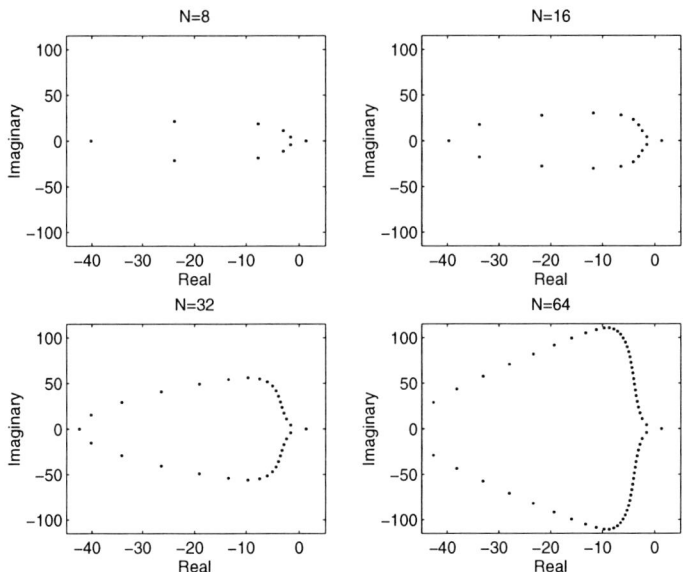

FIG. 4. Eigenvalues of $\widehat{A}_{\text{NEW}}^N$ (equivalent inner product)

Here $y(t, \xi)$ represents displacement at time t and position ξ along an elastic string of length 1. This model of an elastic string with boundary damping is well known and has been considered by many authors (see for example [6, 14, 15, 18, 19, 31, 33, 38] and [37, (example 4)]). There are many interesting questions concerning this example, but in keeping with our previous discussion we will only consider the issue of preserving uniform stability under approximation. The velocity feedback term can be considered as a boundary controller and the problem can be treated as an LQR problem, but we will instead treat this energy dissipation term as part of the system dynamics and consider (3.20) as a simulation problem (in [24] the authors *do* consider the LQR problem and use a regularized algebraic Riccati equation to overcome the lack of uniform stability which arises for some approximation schemes for this problem).

By choosing a state $z(t) = (y(l, \cdot), y_t(l, \cdot))$, we can write (3.20) as the Cauchy problem

$$\begin{aligned} \dot{z}(t) &= Az(t) \\ z(0) &= (y_0, v_0) \end{aligned} \quad (3.21)$$

on the Hilbert space $H = H_L^1(0,1) \times L^2(0,1)$, where

$$H_L^1(0,1) = \{u \in H^1(0,1) : u(0) = 0\}.$$

The energy norm on H is given by

$$\|(u,v)\|_H^2 = \int_0^1 (|u'(\xi)|^2 + |v(\xi)|^2)d\xi,$$

with the compatible energy inner product $\langle \cdot, \cdot \rangle_H$. The operator A in (3.21) is defined on the domain

$$\text{dom } A = \{(u,v) \in H : u \in H^2(0,1),\ v \in H_L^1(0,1),\ u'(1) = -\alpha v(1)\}$$

by

$$A(u,v) = (v, u'').$$

It is known (see [14, 31]) that A is the infinitesimal generator of an exponentially stable semigroup $T(t)$. In [6] the authors investigated the uniform stability behavior (or lack thereof) for a number of standard approximation schemes, including finite differences, Galerkin finite elements, mixed methods, spectral methods, etc. Through numerical experiments (i.e., computing the eigenvalues of the finite-dimensional operators A^N arising for each of these schemes) they discovered that many of these schemes *do not* uniformly preserve the exponential stability of the original system. (Subsequent rigorous theoretical verification of some of the disturbing numerical behavior can be found in [35].) This is noteworthy and perhaps even alarming, given that these are popular approximation schemes and this is a seemingly reasonable physical model of a damped elastic system. The reader is referred to [6] for more detailed discussion and examples. Here we will only illustrate typical behavior by considering a standard Galerkin finite element scheme.

Toward this end, define the Hilbert space $V = H_L^1(0,1) \times H_L^1(0,1) \subset H$ with norm

$$\|(u,v)\|_V^2 = \int_0^1 (|u'(\xi)|^2 + |v'(\xi)|^2) d\xi$$

and the sesquilinear form $\sigma : V \times V \to \mathbb{C}$ by

$$\sigma((u,v),(f,g)) = \int_0^1 [v'(\xi)\overline{f'(\xi)} - u'(\xi)\overline{g'(\xi)}]d\xi - \alpha v(1)\overline{g(1)}. \qquad (3.22)$$

Then the operator A is related to σ by

$$\langle A(u,v),(f,g)\rangle_H = \sigma((u,v),(f,g)) \quad \text{for all } (u,v),(f,g) \in V.$$

Next, let $\xi_i^N = i/N$, $i = 0, \ldots, N$ be a partition of the interval $[0,1]$. Define linear splines $h_i^N(\xi)$ by

$$h_i^N(\xi) = \begin{cases} N(\xi - \xi_{i-1}^N) & \text{if } \xi_{i-1}^N \leq \xi \leq \xi_i^N, \\ -N(\xi - \xi_{i+1}^N) & \text{if } \xi_i^N \leq \xi \leq \xi_{i+1}^N, \\ 0 & \text{elsewhere,} \end{cases} \quad i = 1, \ldots, N-1$$

$$h_N^N(\xi) = \begin{cases} N(\xi - \xi_{N-1}^N) & \text{if } \xi_{N-1}^N \leq \xi \leq 1, \\ 0 & \text{elsewhere.} \end{cases}$$

Set $S_1^N = \text{span}\{h_i^N\}_{i=1}^N$ and $V^N = S_1^N \times S_1^N$. Then $V^N \subset V$ and the operator $A^N : V^N \to V^N$ is defined by

$$\langle A^N(u,v),(f,g)\rangle_H = \sigma((u,v),(f,g)) \quad \text{for all } (u,v),(f,g) \in V^N. \qquad (3.23)$$

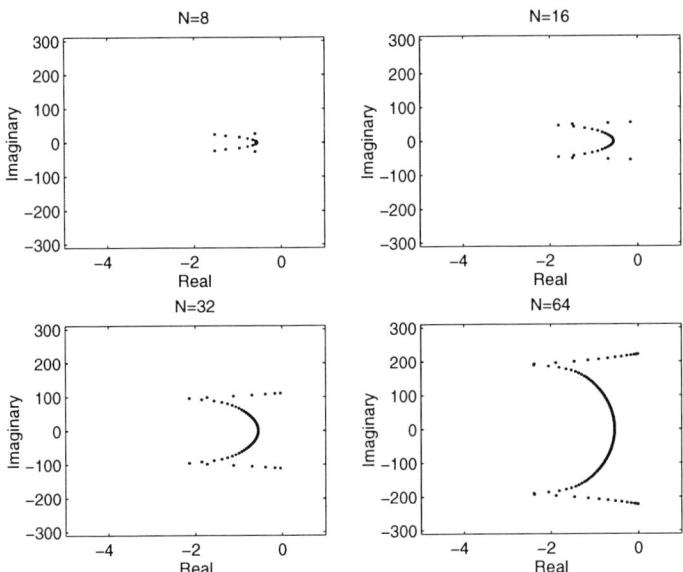

FIG. 5. Eigenvalues of A^N (finite element method)

This is a standard Galerkin finite element approximation scheme for (3.20). Even though $T(t)$ is exponentially stable, this scheme is not uniformly exponentially stable. This is seen convincingly in Fig. 5, in which we plot (for the case $\alpha = 0.5$ in the boundary condition) the eigenvalues of A^N for $N = 8, 16, 32, 64$ (note: for scaling purposes a few eigenvalues with large negative real part have been omitted from this figure). This is typical of the behavior observed by the authors of [6] for several different schemes. In particular, the popular finite difference scheme suffers a similar behavior for its eigenvalues. In [6] the authors used mixed element methods to construct some uniformly exponentially stable schemes.

It is also possible to define Galerkin projections in a different inner product, and this has been pursued in [19]. In particular, define the inner product $\langle \cdot, \cdot \rangle_e$ by

$$\langle (u,v), (f,g) \rangle_e = \langle (u,v), (f,g) \rangle_H + \int_0^1 a(\xi)[u'(\xi)\overline{g(\xi)} + v(\xi)\overline{f'(\xi)}]d\xi, \quad (3.24)$$

where $a(\xi) = e^{\gamma \xi} - 1$ and $\gamma = \ln \frac{(1+\alpha)^2}{1+\alpha^2}$. Also set $V_e = H^2(0,1) \cap H_L^1(0,1) \times H_L^1(0,1)$ with norm

$$\|(u,v)\|_{V_e}^2 = \int_0^1 (|u''|^2 + |u'|^2 + |v'|^2)d\xi,$$

and define a sesquilinear form $\sigma_e : V_e \times V_e \to \mathbb{C}$ by

$$\sigma_e((u,v),(f,g)) = \int_0^1 [v'(\xi)\overline{f'(\xi)} - u'(\xi)\overline{g'(\xi)}]d\xi - \alpha v(1)\overline{g(1)}$$

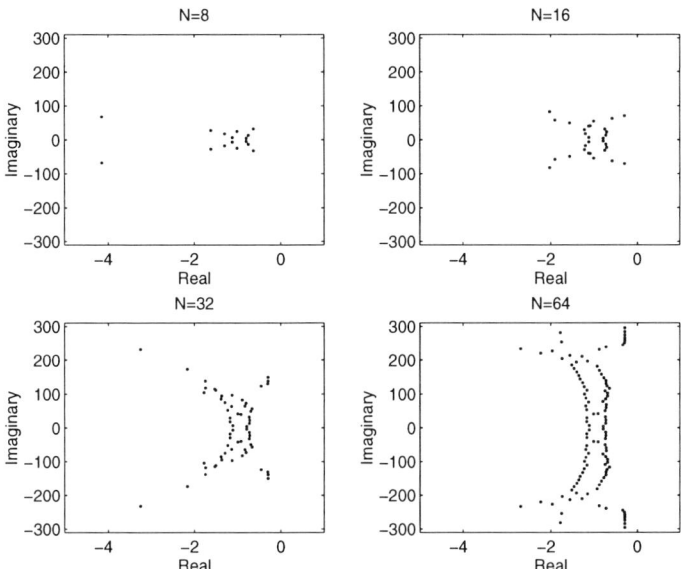

FIG. 6. Eigenvalues of A_e^N (Galerkin method in equivalent inner product)

$$+ \int_0^1 a(\xi)[u''(\xi)\overline{f'(\xi)} + v'(\xi)\overline{g(\xi)}]d\xi$$
$$- a(1)[u'(1) + \alpha v(1)]\overline{f'(1)}.$$

Observe that $\text{dom}\, A \subset V_e$ and that

$$\sigma_e(x,y) = \langle Ax, y \rangle_e \qquad \forall x \in \text{dom}\, A,\ y \in V_e. \tag{3.25}$$

Notice that we cannot define A_e^N by using V_e, σ_e and V^N in (3.23), because $V^N \not\subset V_e$. Instead, set

$$S_2^N = \text{span}\{ \int_0^\xi h_i^N(t)\, dt \}_{i=1}^N,$$

and define $V_e^N = S_2^N \times S_1^N$. Now, $V_e^N \subset V_e$ and we can define $A_e^N : V_e^N \to V_e^N$ by

$$\langle A_e^N x, y \rangle_e = \sigma_e(x,y) \quad \forall x, y \in V_e^N.$$

It can be shown theoretically that the semigroups $T_e^N(t) = e^{tA_e^N}$ are uniformly exponentially stable (see [19] for details). We can see this numerically in Fig. 6, in which the eigenvalues of A_e^N are plotted for $N = 8, 16, 32, 64$.

4 The Parameter Estimation Problem

In this section we continue the theme of investigating computational issues for semidiscrete approximations of infinite-dimensional problems, and in particular we focus on the parameter estimation problem. To begin, consider the following Cauchy problem evolving in a Hilbert space H:

$$\begin{aligned}\dot{z}(t) &= A(q)z(t) + f(t), \quad 0 < t < T, \\ z(0) &= z_0.\end{aligned} \qquad (4.1)$$

Here the operator $A(q)$ depends on a parameter $q \in \mathcal{Q} \subset \mathbb{R}^M$. We assume that that $f \in L_1(0,T;H)$ and that for each $q \in \mathcal{Q}$, the operator $A(q)$ satisfies $\mathrm{dom} A(q) \subset H$ and that $A(q)$ is the infinitesimal generator of a C_0-semigroup $T(t;q)$ on H. We can write (4.1) in the equivalent variation of constants form

$$z(t;q) = T(t;q)z_0 + \int_0^t T(t-s;q)f(s)\,ds \qquad (4.2)$$

for each $q \in \mathcal{Q}$. Roughly speaking, a parameter estimation problem involves estimating the parameter q using data consisting of observations of the system (4.2). There are many ways to formulate such a problem, and we restrict ourselves to consideration of a least-squares fit-to-data formulation, with data consisting of observations of the system at discrete times t_i and discrete spatial locations x_j. That is, we are assuming that $H = L_2(\Omega)$ where Ω is some spatial domain, so that $z(t_i, x_j; q)$ makes sense and represents system output which is to be fit to data d_{ij} which represents actual system observations. We refer the reader to [9] and the references therein for discussion of more general formulations of the parameter estimation problem (for example, one could allow a parameter-dependent forcing term $f(t;q)$ in (4.1), parameter-dependent Hilbert spaces $H(q)$, infinite-dimensional parameter space $\mathcal{Q}$, different data forms, etc.).

We consider the least-squares problem of minimizing the functional

$$J(q) = \sum_{i,j} |z(t_i, x_j; q) - d_{ij}|^2 \qquad (4.3)$$

over $q \in \mathcal{Q}$, subject to (4.2). This is a nonlinear optimization problem with infinite-dimensional constraints (4.2). As before, to solve this problem computationally we need a semidiscrete approximation scheme and appropriate convergence results. Thus, let us introduce a semidiscrete approximation scheme consisting of a sequence of finite-dimensional subspaces $H^N \subset H$, $N = 1, 2, \ldots$, the corresponding orthogonal projections $P^N : H \to H^N$, and operators $A^N(q) : H^N \to H^N$ which generate C_0-semigroups $T^N(t;q) = e^{A^N(q)t}$ on H^N. This defines a finite-dimensional system

$$\begin{aligned}\dot{z}^N(t) &= A^N(q)z^N(t) + P^N f(t), \quad 0 < t < T, \\ z(0) &= P^N z_0,\end{aligned} \qquad (4.4)$$

or, equivalently,

$$z^N(t;q) = T^N(t;q)P^N z_0 + \int_0^t T^N(t-s;q)P^N f(s)\,ds. \qquad (4.5)$$

The approximate least-squares problem is to minimize the functional

$$J^N(q) = \sum_{i,j} |z^N(t_i, x_j; q) - d_{ij}|^2 \tag{4.6}$$

over $q \in \mathcal{Q}$, subject to (4.5).

This is a nonlinear optimization problem with finite-dimensional constraints, and there are many reasonable computational algorithms available for its solution (for example, see [9] and [21]). Such an algorithm will produce an optimal least-squares solution $\bar{q}^N$ for each N, and we desire a convergence result along the lines of $\bar{q}^N \to \bar{q}$, where $\bar{q}$ is an optimal least-squares solution (not necessarily unique) for the original infinite-dimensional problem. Assuming that $\mathcal{Q}$ is compact, one possibility for developing a satisfactory convergence theory ([4, 9, 10]) is based on establishing the following "forward" convergence result:

▷ For arbitrary $\{q^N\}$ in $\mathcal{Q}$ with $q^N \to q$ we have $z^N(t; q^N) \to z(t; q)$ in an appropriate norm (frequently stronger than the H norm).

From our analysis of the simulation problem, we know that for each fixed q, (H1) and (H2) imply $z^N(t; q^N) \to z(t; q)$ in the H norm. In view of the parameter-dependence of the semigroup $T(t; q)$ and the form of the functionals J and J^N (pointwise evaluations of the state), we consider the following hypotheses:

(H7) For each $q \in \mathcal{Q}$ and $z \in H$, $T^N(t; q) P^N z \to T(t; q) z$ in H, uniformly on compact t-intervals.

(H8) The convergence $z^N(t; q) \to z(t; q)$ obtained using (4.2), (4.5), (H1) and (H7) is sufficiently strong so that $z^N(t, x; q) \to z(t, x; q)$ for $t \in (0, T)$, $x \in \Omega$.

There are results available to show that (H1), (H7) and (H8) imply the convergence $\bar{q}^N \to \bar{q}$. This indicates that the strategy "first discretize, then optimize" may be reasonable for the infinite-dimensional least-squares problem under consideration. However, recent computational results (see below) indicate that this is not always the case, and, motivated by our discussion for the optimal control problem, we shall try to provide a possible explanation for this.

Many optimization algorithms (e.g., see [17] and [9, p. 173-175] and the references in each) require the computation of gradients (i.e., $\partial J / \partial q_m$, $m = 1, \ldots, M$). One popular method for computing gradients in optimization problems with differential equation constraints is the so-called costate or adjoint method (see [9, 13]). In this method the gradient is defined in terms of a solution of a costate equation which involves the adjoint operator. In particular (we refer the reader to [9, p. 175-179] for details), one can formally rewrite (4.3) as

$$J(q) = \sum_{i,j} \int_0^T |z(t, x_j; q) - \bar{d}_j(t)|^2 \delta(t - t_i) \, dt, \tag{4.7}$$

where z is defined by (4.1) and $\bar{d}_j(t) = d_{ij}$ for $t_{i-1} < t \le t_i$. One then finds formally that

$$\frac{\partial J}{\partial q_m}(q) = -\int_0^T \left\langle \lambda(t), \frac{\partial A}{\partial q_m}(q) z(t;q) \right\rangle_H dt \tag{4.8}$$

with the costate variable λ defined by

$$\begin{aligned}\dot{\lambda}(t) &= -A^*(q)\lambda(t) + \Gamma(t;q), \quad 0 < t < T, \\ \lambda(T) &= 0,\end{aligned} \tag{4.9}$$

and

$$\Gamma(t,x;q) = 2 \sum_{i,j} \left([z(t,x;q) - \bar{d}_j(t)]\delta(t - t_i)\delta(x - x_j)\right).$$

Once an approximation scheme is given for the original least-squares problem, it defines a natural approximation of the costate equation and the formal costate-dependent gradients. More specifically, we have (tacitly assuming that our approximation scheme is such that $H^N \subset \mathrm{dom}A$)

$$\frac{\partial J^N}{\partial q_m}(q) = -\int_0^T \left\langle \lambda^N(t), \frac{\partial A}{\partial q_m}(q) z^N(t;q) \right\rangle_H dt \tag{4.10}$$

with the approximate costate variable λ^N defined by

$$\begin{aligned}\dot{\lambda}^N(t) &= -A^{*N}(q)\lambda^N(t) + P^N \Gamma^N(t;q), \quad 0 < t < T, \\ \lambda^N(T) &= 0,\end{aligned} \tag{4.11}$$

and

$$\Gamma^N(t;q) = 2 \sum_{i,j} \left([z^N(t;q) - \bar{d}_j(t)]\delta(t - t_i)\delta(x - x_j)\right).$$

Now, if we are using a finite-dimensional optimization algorithm which requires gradients, and if such gradients are calculated *without* making use of the costate variable (using, say, finite differences as in the Levenberg–Marquardt algorithm), then it is reasonable to expect convergence $\bar{q}^N \to q$ with the hypotheses (H1), (H7), and (H8) as discussed above. However, if the gradients are calculated with the costate variable, then we could reasonably argue that such an optimization algorithm defines the optimal solution $\bar{q}^N$ in terms of the costate variable λ^N, which is in turn defined as a solution to an equation involving the adjoint operator. (See [1] for a precise statement of a convergence result for such an approach.) But this is precisely the same scenario that arises in the LQR problem (where the optimal control $\tilde{u}^N$ is defined in terms of the operator Π^N, which is in turn defined as a solution to an algebraic Riccati equation involving the adjoint operator), and we have seen there that an additional hypothesis of adjoint semigroup convergence (H6) is needed to establish theoretical convergence. We have also seen in the LQR problem that approximation schemes (e.g. the SPL scheme) that don't satisfy (H6) may converge for the simulation problem but not for the LQR problem. Our contention is that this is a possible explanation for the poor performance of some parameter estimation approximation schemes which utilize costate-based gradient calculations. Clearly the theoretical issue is still unresolved, but there is numerical evidence to support this convention. For example, in [3] Banks et al. used methods such as those described here to estimate material parameters in models of elastic beams. They found that for models in which the system operator $A(q)$ is skew-adjoint (i.e., $A(q)^* = -A(q)$, as occurs in beams with no damping), the costate-based gradient methods were highly competitive with other methods such as Levenberg–Marquardt. But in models which included damping (so that $A(q)$ is not skew-adjoint), Banks et al. experienced extreme difficulties (fitting the data) with the costate-based gradient methods and abandoned them in favor of the Levenberg–Marquardt algorithm. Of course, for skew-adjoint system operators, strong semigroup convergence guarantees strong adjoint semigroup convergence, but this is not necessarily true for non-skew-adjoint system operators.

Acknowledgements

The research of the first author was supported in part by the Air Force Office of Scientific Research under grant AFOSR F4962095-1-0236. The research of the second author was supported in part by the National Science Foundation under grant DMS-9696239 and UNCG New Faculty grant 97NF04.

Bibliography

1. Banks, H. T. (1992). Computational issues in parameter estimation and feedback control problems for partial differential equation systems. *Physica D*, **60**, 226–238.
2. Banks, H. T. and Burns, J. A. (1978). Hereditary control problems: numerical methods based on averaging approximations. *SIAM J. Control and Optimization*, **16**, 169–208.
3. Banks, H. T., Crowley, J. M., and Rosen, I. G. (1986). Methods for the identification of material parameters in distributed models for flexible structures. *Mat. Aplicada e Computational*, **5**, 139–168. ICASE Report #84-66, NASA Langley Research Center.
4. Banks, H. T. and Ito, K. (1988). A unified framework for approximation in inverse problems for distributed parameter systems. *Control: Theory and Advanced Technology*, **4**, 73–90.
5. Banks, H. T., Ito, K., and Rosen, I. G. (1984). A spline based technique for computing Riccati operators and feedback controls in regulator problems for delay equations. *SIAM J. Scientific and Statistical Computing*, **5**, 830–855.
6. Banks, H. T., Ito, K., and Wang, C. (1991). Exponentially stable approximations of weakly damped wave equations. In: *International Series in Numerical Mathematics*, vol. 100, pp. 1–33. Birkhäuser.

7. Banks, H. T. and Kappel, F. (1979). Spline approximations for functional differential equations. *Journal of Differential Equations*, **34**, 496–522.
8. Banks, H. T. and Kunisch, K. (1984). The linear regulator problem for parabolic systems. *SIAM J. Control and Optimization*, **22**, 684–699.
9. Banks, H. T. and Kunisch, K. (1989). *Estimation Techniques for Distributed Parameter Systems*. Birkhäuser, Boston.
10. Banks, H. T. and Rebnord, D. A. (1991). Analytic semigroups: Applications to inverse problems for flexible structures. In: *Differential Equations with Applications* J. Goldstein et al. (eds), pp. 21–35. Dekker.
11. Banks, H. T., Smith, R. C., and Wang, Y. (1996). *Smart Material Structures: Modeling, Estimation and Control*. Masson/Wiley, Paris/Chichester.
12. Burns, J. A., Ito, K., and Propst, G. (1988). On nonconvergence of adjoint semigroups for control systems with delays. *SIAM J. Control and Optimization*, **26**, 1442–1454.
13. Chavent, G. (1979). Identification of DPS: about the output least-squares method, its implimentation, and identifiability. In: *Proc. 5th IFAC Symposium on Identification and System Parameter Estimation*, pp. 85–97. Pergamon.
14. Chen, G. (1979). Energy decay estimates and exact boundary value controllability for the wave equation in a bounded domain. *Math. Pures Appl.*, **58**, 249–274.
15. Cox, S. and Zuazua, E. (1995). The rate at which energy decays in a string damped at one end. *Indiana Univ. Math. J.*, **44**, 545–573.
16. Curtain, R. F. and Zwart, H. J. (1995). *An Introduction to Infinite-Dimensional Linear Systems Theory*, vol. 21 of Texts in Applied Mathematics. Springer.
17. Dennis, J. E. and Schnabel, R. B. (1983). *Numerical Methods for Unconstrained Optimization and Nonlinear Problems*. Prentice Hall.
18. Driscoll, T. A. and Trefethen, L. N. (1996). Pseudospectra for the wave equation with an absorbing boundary. *J. Comp. and Applied Math.*, **69**, 125–142.
19. Fabiano, R. H. (1996). Stability preserving Galerkin approximations for linear distributed parameter systems. Preprint.
20. Fabiano, R. H. (1995). Stability preserving spline approximations for scalar functional differential equations. *Computers Math. Applic.*, **29**, 87–94.
21. Fletcher, R. (1980). *Practical Methods of Optimization*. Wiley.
22. Gibson, J. S. (1979). The Riccati's integral equations for optimal control problems in Hilbert spaces. *SIAM J. Control and Optimization*, **17**, 537–565.
23. Gibson, J. S. (1983). Linear-quadratic optimal control of hereditary differential systems: Infinite-dimensional Riccati equations and numerical approx-

imations. *SIAM J. Control and Optimization*, **21**, 95–139.

24. Hendrickson, E. and Lasiecka, I. (1994). Numerical approximations and regularizations of Riccati equations arising in hyperbolic dynamics with unbounded control operators. *Computational Optimization and Applications*, **2**, 343–390.

25. Ito, K. (1990). Finite-dimensional compensators for infinite-dimensional systems via Galerkin-type approximation. *SIAM J. Control and Optimization*, **28**, 1251–1269.

26. Ito, K. and Kappel, F. (1991). On variational formulations of the Trotter-Kato theorem. Technical report, CAMS #91-7. Center for Applied Mathematical Sciences, University of Southern California.

27. Ito, K. and Kappel, F. (1992). Two families of approximation schemes for delay systems. *Results in Mathematics*, **21**, 93–137.

28. Kappel, F. and Salamon, D. (1987). Spline approximation for retarded systems and the Riccati equation. *SIAM J. Control and Optimization*, **25**, 1082–1117.

29. Kappel, F. and Salamon, D. (1989). On the stability properties of spline approximations for retarded systems. *SIAM J. Control and Optimization*, **27**, 407–431.

30. Kappel, F. and Salamon, D. (1990). An approximation theorem for the algebraic Riccati equation. *SIAM J. Control and Optimization*, **28**, 1136–1147.

31. Lagnese, J. (1983). Decay of solutions of wave equations in a bounded region with boundary dissipation. *Journal of Differential Equations*, **50**, 163–182.

32. Lax, P. D. and Richtmyer, R. D. (1956). Survey of the stability of linear finite difference equations. *Comm. Pure Appl. Math.*, **9**, 267–293.

33. Majda, A. (1975). Disappearing solutions for the dissipative wave equation. *Indiana Univ. Math. J.*, **24**, 1119–1133.

34. Pazy, A. (1983). *Semigroups of Linear Operators and Applications to Partial Differential Equations*, vol. 44 of Applied Mathematical Sciences. Springer.

35. Peichl, G. and Wang, C. (1997). Asymptotic analysis of stabilizability of a control system for a discretized boundary damped wave equation. To appear.

36. Salamon, D. (1985). Structure and stability of finite-dimensional approximations for functional differential equations. *SIAM J. Control and Optimization*, **23**, 928–951.

37. Trefethen, L. N. (1997). Pseudospectra of linear operators. In: *ICIAM '95: Proceedings of the Third International Congress on Industrial and Applied Mathematics*. Akademie Verlag. *SIAM Rev.*, **39**, 383–406.

38. Wang, C. (1995). Linear quadratic optimal control of a wave equation with boundary damping and pointwise control input. *Journal of Mathematical Analysis and Applications*, **192**, 562–578.

8
Zero Distribution, the Szegő Curve, and Weighted Polynomial Approximation in the Complex Plane

I. E. Pritsker and R. S. Varga
Kent State University

1 Introduction

Given $f(z) := \sum_{k=0}^{\infty} a_k z^k$ with $\overline{\lim_{k \to \infty}} |a_k|^{1/k} = 1$, so that f is an analytic function in $|z| < 1$, it is natural to ask where the zeros of the partial sums of f, namely,

$$s_n(z; f) := \sum_{k=0}^{n} a_k z^k \quad (n \in \mathbb{N}),$$

are located. (Here, $\mathbb{N}$ and $\mathbb{N}_0$ denote, respectively, the set of positive integers and the set of nonnegative integers.) It was shown in 1917 by Jentzsch [6] that *each* z of $|z| = 1$ is an accumulation point of the zeros of $\{s_n(z; f)\}_{n \in \mathbb{N}}$. (A sharper form of this can be found in Szegő [15].) But for an entire function, the behavior of the zeros of its partial sums is quite different. In 1924, Szegő [16] made a substantial first step in this area by considering the special entire function e^z and its familiar partial sums,

$$s_n(z) := \sum_{k=0}^{n} z^k / k! \quad (n \in \mathbb{N}_0).$$

Specifically, an application of the Eneström–Kakeya Theorem (cf. Marden [9, p. 137, Ex. 2]) to the above $s_n(z)$ shows that, for each $n \in \mathbb{N}$, all zeros of $s_n(z)$ lie in the disk $|z| \leq n$. Calling $s_n(nz)$ the *normalized* partial sum, this implies that

$$\text{all zeros of } \{s_n(nz)\}_{n \in \mathbb{N}} \text{ lie in } |z| \leq 1. \quad (1.1)$$

This is clearly indicated in Fig. 1. Consequently, the infinite set of all zeros of $\{s_n(nz)\}_{n \in \mathbb{N}}$ must have at least one accumulation point in $|z| \leq 1$. On defining the simple closed curve

$$S := \{z \in \mathbb{C} : |ze^{1-z}| = 1 \text{ and } |z| \leq 1\}, \quad (1.2)$$

which is called the *Szegő curve*, we next state the remarkable result of Szegő [16], published in 1924.

Theorem 1.1 [16]. *A complex number ζ is an accumulation point of the zeros of the normalized partial sums $\{s_n(nz)\}_{n\in\mathbb{N}}$ if and only if $\zeta \in S$.*

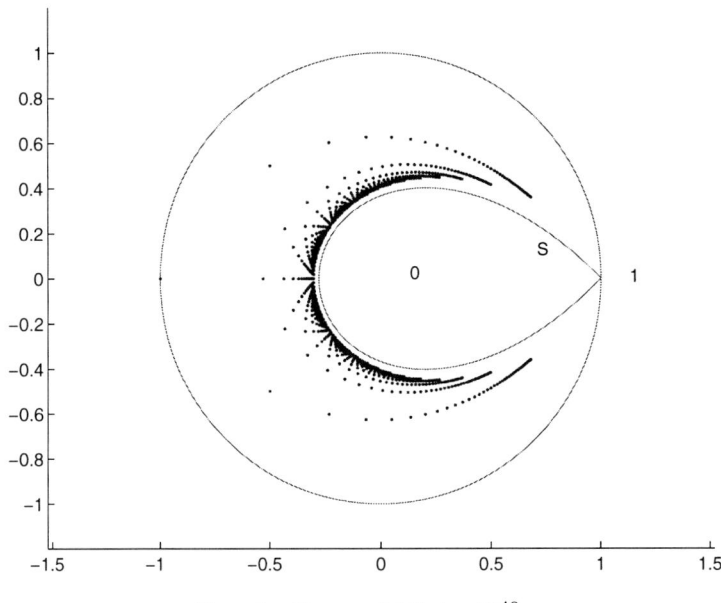

FIG. 1. Zeros of $\{S_n(nz)\}_{n=1}^{40}$

We remark that the Szegő curve S is also shown in Fig. 1. In addition, we note from Fig. 1 that the convergence, of the zeros of $\{s_n(nz)\}_{n\in\mathbb{N}}$ to S, is noticeably *slower* in the neighborhood of the point $z = 1$ of S, and this will have subsequent interesting ramifications for us! We also mention that the curve

$$\{z \in \mathbb{C} : |ze^{1-z}| = 1\}, \tag{1.3}$$

of which the Szegő curve S of (1.2) is a part, is shown in Fig. 2, and this divides the complex plane into three disjoint domains, i.e., G, Ω_0, and Ω_∞, where G denotes the interior of the Szegő curve S. (The significance of these three domains, namely, G, Ω_0, and Ω_∞ will be clarified later.)

2 Weighted Polynomial Approximation by $\{e^{-nz}P_n(z)\}_{n\in\mathbb{N}_0}$

The introduction in Section 1 to the zeros of the normalized partial sums $s_n(nz)$ of e^z may seem remote from our goal of considering weighted polynomial approximation of analytic function in the complex plane, but, as was shown in [16], there holds

$$e^{-nz}s_n(nz) = 1 - \frac{\sqrt{n}}{\tau_n\sqrt{2\pi}} \int_0^z (\zeta e^{1-\zeta})^n d\zeta \quad (n \in \mathbb{N}, z \in \mathbb{C}), \tag{2.1}$$

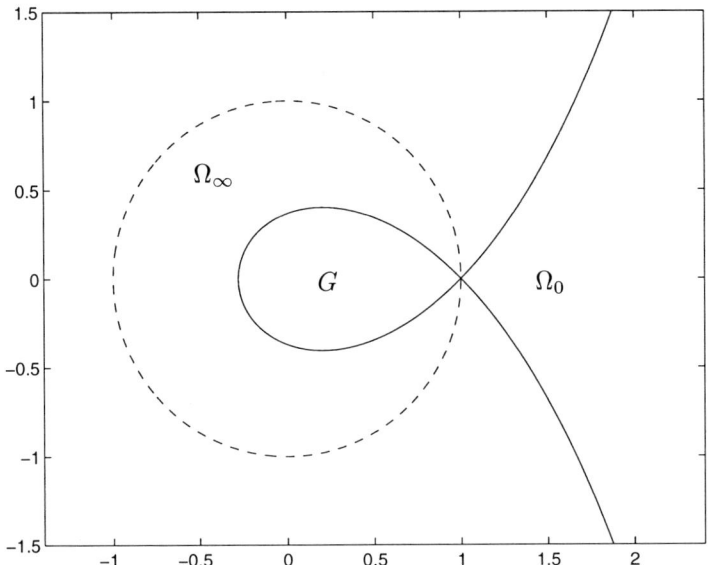

FIG. 2. *The Szegő curve and the associated domains*

where, from Stirling's asymptotic series formula (cf. Henrici [5, p. 377]),

$$\begin{cases} \tau_n := \dfrac{n!}{n^n e^{-n}\sqrt{2\pi n}} \approx 1 + \dfrac{1}{12n} + \dfrac{1}{288n^2} - \dfrac{139}{51840n^3} + \cdots, \text{ as } n \to \infty, \\ \text{so that } \lim_{n \to \infty} \tau_n = 1. \end{cases} \quad (2.2)$$

We note that the function

$$\varphi(z) := ze^{1-z} \quad (2.3)$$

already appears in the integrand of (2.1), and it also plays a significant role in the asymptotic behavior of the zeros of $\{s_n(nz)\}_{n \in \mathbb{N}}$. It is the case that $\varphi(z)$ of (2.3) is univalent in $|z| < 1$, and $\varphi(z) = w$ conformally maps the open set G onto the open unit disk $\{w \in \mathbb{C} : |w| < 1\}$ in the w-plane, with the Szegő curve S being mapped 1–1 onto the unit circle $T := \{w \in \mathbb{C} : |w| = 1\}$. But note also that $\varphi'(1) = 0$, so that conformality of this mapping is lost at the point $z = 1$.

Based on (2.1), the following can be established.

Proposition 2.1 [11] *For the weighted normalized partial sums $e^{-nz} s_n(nz)$, there holds*

$$|e^{nz} s_n(nz) - 1| \leq \dfrac{4}{\sqrt{2\pi n}|z - 1|} \quad (z \in \overline{G} \setminus \{1\}; n \in \mathbb{N}). \quad (2.4)$$

We remark that the inequality of (2.4) of Proposition 2.1, which is in a form useful for our subsequent developments, is closely related to similar results of Szegő [16, eq. (5′)], eq. and [2, eq. (2.13)], but is not implied by these results.

For the behavior of the weighted normalized partial sums $\{e^{-nz}s_n(nz)\}_{n\in\mathbb{N}}$ outside of $\overline{G}$, we also have the following result:

Theorem 2.2 [11] *For the normalized partial sums of e^z, we have*

$$\lim_{n\to\infty} |e^{-nz}s_n(nz)|^{1/n} = |\varphi(z)| \quad (z \in \mathbb{C}\backslash\overline{G}). \tag{2.5}$$

The result (2.5) shows that $e^{-nz}s_n(nz)$ diverges unboundedly in Ω_∞ (where $|\varphi(z)| > 1$), and converges to the identically zero function in Ω_0 (where $|\varphi(z)| < 1$), as the subscripts of these regions indicate. What (2.4) asserts is that the special analytic function, $f(z) \equiv 1$, can be uniformly approximated, on compact subsets of $\overline{G}\backslash\{1\}$, by the weighted normalized partial sums $\{e^{-nz}s_n(nz)\}_{n\in\mathbb{N}}$. This then connects with the theme given in the title of this paper! Moreover, Proposition 2.1 is the precursor of the following more general approximation result for the weighted polynomials $\{e^{-nz}P_n(z)\}_{n\in\mathbb{N}_0}$. In what follows, all norms used are the uniform (Chebyshev) norms on the indicated sets.

Theorem 2.3 [11] *Let f be analytic in G and continuous on compact subsets of $\overline{G}\backslash\{1\}$. Then, given any compact set $E \subset \overline{G}\backslash\{1\}$, there exists a sequence of complex polynomials $\{P_n(z)\}_{n\in\mathbb{N}_0}$, with $\deg P_n \leq n$ for all $n \in \mathbb{N}_0$, such that*

$$\lim_{n\to\infty} \|e^{-nz}P_n(z) - f(z)\|_E = 0. \tag{2.6}$$

Furthermore, if f is analytic in G, and continuous in $\overline{G}$ with $f(1) = 0$, then there is a sequence of polynomials $\{\tilde{P}(z)\}_{n\in\mathbb{N}_0}$, such that

$$\lim_{n\to\infty} \|e^{-nz}\tilde{P}_n(z) - f(z)\|_{\overline{G}} = 0. \tag{2.7}$$

The difference between (2.6) and (2.7) is in the behavior of f at the sole point $z = 1$ of $\overline{G}$, the point which is exceptional in Proposition 2.1, and for which we know that $\varphi'(1) = 0$. We believe that the second part of Theorem 2.3 cannot be further strengthened, in the sense that we make the following:

Conjecture [11] There exists an f, analytic in G and continuous in $\overline{G}$ with $f(1) \neq 0$, such that for *no* sequence of polynomials $\{P_n(z)\}_{n\in\mathbb{N}_0}$, with $\deg P_n \leq n$ for each $n \in \mathbb{N}_0$, it is true that

$$\lim_{n\to\infty} \|e^{-nz}P_n(z) - f(z)\|_{\overline{G}} = 0. \tag{2.8}$$

Let us examine Theorem 2.3 more closely. Obviously, any polynomial $Q_n(z)$ can itself be locally uniformly approximated (i.e., on compact subsets of $\overline{G}\backslash\{1\}$) by weighted polynomials $\{e^{-nz}P_n(z)\}_{n\in\mathbb{N}_0}$. Then, by Mergelyan's Theorem (cf.

Walsh [19, p. 367]), any function which is analytic interior to a given compact set $E \subset \overline{G}\backslash\{1\}$ and is continuous on E, is itself uniformly approximable by polynomials, if E has a connected complement. This gives us the result:

Corollary 2.4 [11] *If E is a compact set, such that $E \subset \overline{G}\backslash\{1\}$ and such that its complement $\overline{\mathbb{C}}\backslash E$ is connected, then there exists a sequence of polynomials $\{P_n(z)\}_{n\in\mathbb{N}_0}$, with $\deg P_n \leq n$, such that*

$$\lim_{n\to\infty} \|e^{-nz} P_n(z) - f(z)\|_E = 0. \tag{2.9}$$

Many questions naturally arise from Theorem 2.3. For example, one can ask if it is possible to give *rates of convergence* in (2.6), of these weighted polynomial approximations to a given f. This is done below, and is reminiscent of the classical Bernstein–Walsh overconvergence results (cf. [19, pp. 75-78]). To state this result, we use the notation

$$S_r := \{z \in \mathbb{C} : |\varphi(z)| = |ze^{1-z}| = r, |z| \leq 1 \text{ and } 0 < r \leq 1\}, \tag{2.10}$$

so that S_r is a *level curve* of the mapping φ. We also set

$$G_r := \text{int } S_r \quad (0 < r \leq 1), \tag{2.11}$$

so that $G = G_1$.

Theorem 2.5 [11] *Let (r, R) be a pair of numbers satisfying $0 < r < R \leq 1$. Then, a function f is analytic in G_R if and only if there exists a sequence of polynomials $\{P_n(z)\}_{n\in\mathbb{N}_0}$, with $\deg P_n \leq n$, such that*

$$\varlimsup_{n\to\infty} \|e^{-nz} P_n(z) - f(z)\|_{S_r}^{1/n} \leq \frac{r}{R} < 1. \tag{2.12}$$

We remark that the last inequality in (2.12) implies the *geometric convergence* of the sequence $\{e^{-nz}P_n(z)\}_{n\in\mathbb{N}}$ to $f(z)$ in S_r, for $0 < r < R \leq 1$.

It also turns out that the case of equality in (2.12) can be studied more carefully. For notation, if f is analytic in G_R, where $0 < r < R < 1$, set

$$E_n^{\exp(-z)}(f, \overline{G}_r) := \inf_{P_n \in \pi_n} \|e^{-nz} P_n(z) - f(z)\|_{S_r}, \tag{2.13}$$

where π_n denotes the collection of all complex polynomials of degree at most n. Thus, $E_n^{\exp(-z)}(f, \overline{G}_r)$ is the best uniform weighted polynomial approximation, from π_n, of f in $\overline{G}_r$, with the weight function $W(z) := e^{-z}$, for each $n \in \mathbb{N}_0$. We then have:

Corollary 2.6 [11] *The function $f(z)$, which is analytic in G_R, has a singularity on S_R if and only if*

$$\varlimsup_{n\to\infty} \{E_n^{\exp(-z)}(f, \overline{G}_r)\}^{1/n} = \frac{r}{R}. \tag{2.14}$$

We next consider the possibility of uniform approximability of functions by $\{e^{-nz}P_n(z)\}_{n\in\mathbb{N}_0}$, in sets *other* than those such as compact subsets of $\overline{G}\setminus\{1\}$, or $\overline{G}$, already treated in Theorem 2.3. This includes the following *shift-invariant property* of weighted polynomial approximation, when the weight function is $W(z) = e^{-z}$.

Theorem 2.7 [11] *Let E be a compact set in $\mathbb{C}$ which has the property that any function, analytic in E and continuous on E, can be uniformly approximated on E by the weighted polynomials $\{e^{-nz}P_n(z)\}_{n\in\mathbb{N}_0}$. Then, the same is true for the translated set $\zeta + E := \{\zeta + z : z \in E\}$, for **any** $\zeta \in \mathbb{C}$.*

Using the above idea, we give the following negative result which shows, up to translations (as considered in the statement of Theorem 2.7), that the open set G, the interior of the Szegő curve S, is the largest *universal* domain of this shape in the complex plane for which locally uniform weighted polynomial approximation, of the form $\{e^{-nz}P_n(z)\}_{n\in\mathbb{N}_0}$, to arbitrary analytic functions is possible.

Theorem 2.8 [11]. *If a domain H, or any of its translations, contains $\overline{G}$, then no analytic function in H (except the identically zero function) can be approximated, locally uniformly in H, by weighted polynomials $\{e^{-nz}P_n(z)\}_{n\in\mathbb{N}_0}$.*

We next show that the weighted polynomials $\{e^{-nz}P_n(z)\}_{n\in\mathbb{N}_0}$ can give rise to locally uniform approximation of analytic functions in domains having shapes *different* from G or its translations. As an example, consider an open disk in $\mathbb{C}$ having a radius r, where $r > 0$. Our next result incorporates the shift invariant property of the weighted polynomials $\{e^{-nz}P_n(z)\}_{n\in\mathbb{N}_0}$.

Theorem 2.9 [11]. *Any function, analytic in an open disk of radius $1/2$, can be approximated by weighted polynomials $\{e^{-nz}P_n(z)\}_{n\in\mathbb{N}_0}$, uniformly on compact subsets of this disk. Furthermore, if H is a domain containing a closed disk of radius $1/2$, then no analytic function (except the identically zero function) in H can be approximated locally uniformly in H by such weighted polynomials $\{e^{-nz}P_n\}_{n\in\mathbb{N}_0}$.*

We note that Theorem 2.9 determines the *largest* disk in which such weighted polynomial approximation, to arbitrary analytic functions, is locally uniformly convergent. In Fig. 3, we similarly show, as in the case of Fig. 2, the case $r = 1/2$ of the curve, defined by

$$\{z \in \mathbb{C} : \log \frac{|z|}{r} + \left(\frac{r^2}{|z|^2} - 1\right) \cdot \operatorname{Re} z = 0, \text{ and } |z| \geq r\}, \qquad (2.15)$$

which divides the complex plane into three disjoint domains, $|z| < 1/2, \tilde{\Omega}_0$, and $\tilde{\Omega}_\infty$. These three domains play exactly the analogous roles, in the convergence of $\{e^{-nz}P_n(z)\}_{n\in\mathbb{N}_0}$, as did G, Ω_0, and Ω_∞ for $\{e^{-nz}s_n(nz)\}_{n\in\mathbb{N}_0}$, (cf. Proposition 2.1 and Theorem 2.2.)

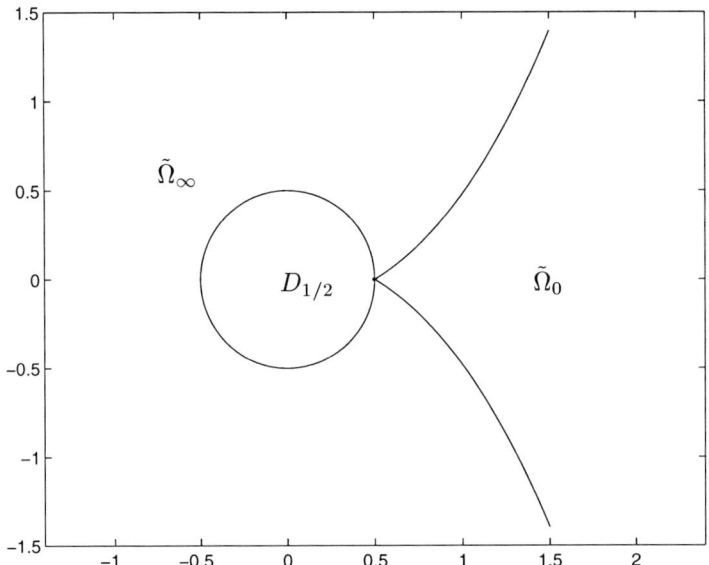

FIG. 3. The extremal disk $D_{1/2}$ and the associated domains

3 General Weighted Polynomial Approximation

In the previous section, we have seen how weighted polynomials, of the form $\{e^{-nz}P_n(z)\}_{n\in\mathbb{N}_0}$, stemming from the work of Szegő [16], can locally uniformly approximate analytic functions in the domain G, defined as the interior of the Szegő curve S of (1.2), or in any of its translations, as well as in the open disk $\{z \in \mathbb{C} : |z| < 1/2\}$, or in any of its translations. But, the broader theoretical question that can be investigated is this. Given the pair

$$(G, W), \tag{3.1}$$

where

$$\begin{cases} \text{i)} & G \text{ is an open bounded set, in the complex plane } \mathbb{C}, \text{ which} \\ & \text{can be represented as a finite or countable union of disjoint} \\ & \text{simply connected domains, i.e., } G = \bigcup_{\ell=1}^{\sigma} G_\ell \text{ (where } 1 \leq \sigma \leq \infty \text{);} \\ \text{ii)} & W(z), \text{ the weight function, is analytic in } G \text{ with } W(z) \neq 0 \\ & \text{for any } z \in G, \end{cases} \tag{3.2}$$

when is it the case that this pair (G, W) has the **approximation property**, i.e.,

$$\begin{cases} \text{for any } f(z) \text{ which is analytic in } G \text{ and for any compact subset} \\ E \text{ of } G, \text{ there exists a sequence of polynomials } \{P_n(z)\}_{n=0}^{\infty}, \\ \text{with } \deg P_n \leq n \text{ for all } n \in \mathbb{N}_0, \text{ such that} \\ \qquad \lim_{n \to \infty} \|f - W^n P_n\|_E = 0. \end{cases} \quad (3.3)$$

The material of Section 2 corresponded to the specific choice of the weight function $W(z) := e^{-z}$.

Given a pair (G, W), as in (3.1), we state below our main result, Theorem 3.1, which gives a characterization, in terms of potential theory, for the pair (G, W) to have the approximation property. For notation, let $\mathcal{M}(E)$ be the space of all positive unit Borel measures on $\mathbb{C}$ which are supported on a compact set E, i.e., for any $\mu \in \mathcal{M}(E)$, we have $\mu(\mathbb{C}) = 1$ and $\operatorname{supp} \mu \subset E$. The logarithmic potential of a compactly supported measure μ is then defined (cf. Tsuji [18, p. 53]) by

$$U^\mu(z) := \int \log \frac{1}{|z-t|}\, d\mu(t). \quad (3.4)$$

Theorem 3.1 [12] *A pair (G, W), as in (3.1), has the approximation property (3.3) if and only if there exist a measure $\mu(G, W) \in \mathcal{M}(\partial G)$ and a constant $F(G, W)$ such that*

$$U^{\mu(G,W)}(z) - \log|W(z)| = F(G, W), \quad \text{for any } z \in G. \quad (3.5)$$

Remark It is well known that any open set in the complex plane is a finite or countable union of disjoint domains, and this is more general than the assumption on the open set G in (3.2(i)). However, we note that the approximation property (3.3) *cannot hold*, even in the classical case where $W(z) \equiv 1$ for all $z \in G$, if $G = \bigcup_{\ell=1}^{\sigma} G_\ell$, when some G_ℓ is multiply connected (cf. Walsh [19, p. 25]). In this sense, our initial assumptions on G in (3.2(i)) are quite general.

Remark The condition that $W(z) \neq 0$ for all $z \in G$ cannot be dropped, for if $W(z_0) = 0$ for some $z_0 \in G_\ell$, where $G = \bigcup_{\ell=1}^{\sigma} G_\ell$, then the necessarily null sequence $\{W^n(z_0) P_n(z_0)\}_{n=0}^{\infty}$ trivially fails to converge to any $f(z)$, analytic in G, with $f(z_0) \neq 0$; whence, the approximation property fails. Even more decisive is the result, to be proved in Section 5, that if $W(z_0) = 0$ for some $z_0 \in G_\ell$, then the sequence $\{W^n(z) P_n(z)\}_{n=0}^{\infty}$ can converge, locally uniformly in G, to $f(z)$, only if $f(z) \equiv 0$ in G_ℓ. In this sense, the assumptions on $W(z)$ in (3.2(ii)) are also quite general.

Remark In the case $W(z) \equiv 1$ of Theorem 3.1, the result that the approximation property (3.3) holds is a known classical result in complex approximation theory (cf. [19, p. 26]). This also follows from Theorem 3.1 because the measure $\mu(G, 1)$ exists by Theorems III.12 and III.14 of Tsuji [18], and is the classical equilibrium distribution measure (in the sense of logarithmic potential theory) for $\overline{G}$.

The topic of weighted approximation by $\{W^n(z)P_n(z)\}_{n=0}^\infty$, on the real line, has been extensively and thoroughly treated in the recent books of Saff and Totik [13] and Totik [17]. Here, we emphasize weighted approximation *in the complex plane*, which has received far less attention in the current approximation theory literature, with the exception of the recent papers by Borwein and Chen [1] and Pritsker and Varga [11, 12].

We shall present in Section 4 a number of applications of Theorem 3.1 to special pairs (G,W). The proofs of some results and remarks on weighted approximation, stated in Sections 3 and 4, are given in Section 5. Finally, we conclude this chapter with Section 6, where further remarks, open problems, and a discussion of possible generalizations are given.

4 Applications

Finding the measure $\mu(G,W)$ of Theorem 3.1 or verifying its existence is a nontrivial problem in general. Since $U^{\mu(G,W)}(z)$ in (3.5) is harmonic in $\mathbb{C}\setminus \operatorname{supp}\mu(G,W)$ and, since it can be shown from (3.5), if $\log|W(z)|$ is continuous on $\overline{G}$ and if G is a *finite* union of $G_\ell, \ell = 1, 2, \ldots, \ell_0$, that $U^{\mu(G,W)}(z)$ is equal to $\log|W(z)| + F(G,W)$ on $\operatorname{supp}\mu(G,W)$, then $U^{\mu(G,W)}(z)$ can be found as the solution of the corresponding Dirichlet problems. The measure $\mu(G,W)$ can be recovered from its potential, using the Fourier method described in Section IV.2 of Saff and Totik [13]. This method has already been used successfully by the authors in [11] to study the approximation of analytic functions by the weighted polynomials $\{e^{-nz}P_n(z)\}_{n=0}^\infty$, i.e., when $W(z) := e^{-z}$.

In contrast to the above procedure, we next consider a different method, dealing with specific weight functions, which allows us to deduce "explicit" expressions for the measure $\mu(G,W)$ of Theorem 3.1, and to treat some important cases of pairs (G,W). With G, as defined in (3.2(i)) with σ finite, we denote the unbounded component of $\overline{\mathbb{C}}\setminus\overline{G}$ by Ω. Let ν_1 and ν_2 be two unit positive Borel measures on $\mathbb{C}$ with compact supports satisfying

$$\operatorname{supp}\nu_1 \subset \overline{\mathbb{C}}\setminus G \quad \text{and} \quad \operatorname{supp}\nu_2 \subset \overline{\mathbb{C}}\setminus G, \tag{4.1}$$

such that

$$\nu_1(\mathbb{C}) = \nu_2(\mathbb{C}) = 1. \tag{4.2}$$

For real numbers α and β, assume that $W(z)$, satisfying

$$\log|W(z)| = -\left(\alpha U^{\nu_1}(z) + \beta U^{\nu_2}(z)\right), \quad z \in G, \tag{4.3}$$

is analytic in G. Then, we state, as an application of Theorem 3.1, our next result as

Theorem 4.1 [12] *Given any pair of real numbers α and β, given an open bounded set $G = \bigcup_{\ell=1}^\sigma G_\ell$ as in (3.2(i)) with σ finite, and given the weight function $W(z)$ of (4.3), then the pair (G,W) has the approximation property (3.3) if and only if the measure*

$$\mu := (1 + \alpha + \beta)\omega(\infty, \cdot, \Omega) - \alpha \hat{\nu}_1 - \beta \hat{\nu}_2 \qquad (4.4)$$

is positive, where $\omega(\infty, \cdot, \Omega)$ is the harmonic measure at ∞ with respect to Ω; here, $\hat{\nu}_1$ and $\hat{\nu}_2$ are, respectively, the balayages of ν_1 and ν_2 from $\overline{\mathbb{C}} \backslash \overline{G}$ to $\overline{G}$. Furthermore, if μ of (4.4) is a positive measure, then (cf. Theorem 3.1)

$$\mu(G, W) = \mu \quad \text{and} \quad \operatorname{supp} \mu(G, W) \subset \partial G. \qquad (4.5)$$

We point out that the harmonic measure $\omega(\infty, \cdot, \Omega)$ (cf. Nevanlinna [10] and Tsuji [18]) is the same as the equilibrium distribution measure for $\overline{G}$, in the sense of classical logarithmic potential theory [18]. For the notion of balayage of a measure, we refer the reader to Chapter IV of Landkof [7] or Section II.4 of Saff and Totik [13].

In the following series of subsections, we consider various classical weight functions and find their corresponding measures, associated with the weighted approximation problem in G by Theorem 3.1.

4.1 Incomplete Polynomials and Laurent Polynomials

The *incomplete polynomials* of Lorentz [8] are a sequence of polynomials of the form

$$\left\{ z^{m(i)} P_{n(i)}(z) \right\}_{i=0}^{\infty}, \quad \deg P_{n(i)} \leq n(i), \ (m(i), n(i) \in \mathbb{N}_0), \qquad (4.6)$$

where it is assumed that $\lim_{i \to \infty} \dfrac{m(i)}{n(i)} =: \alpha$, where $\alpha > 0$ is a real number. The question of the possibility of approximation by incomplete polynomials is closely connected to that of approximation by the weighted polynomials

$$\{z^{\alpha n} P_n(z)\}_{n=0}^{\infty}, \quad \deg P_n \leq n. \qquad (4.7)$$

The question of approximation by the incomplete polynomials of (4.6) was completely settled by Saff and Varga [14], and by Golitschek [4] on the interval $[0,1]$ (see Totik [17] and Saff and Totik [13] for the associated history and later developments). We consider now the analogous problem in the complex plane. Since the weight $W(z) := z^\alpha$ in (4.7) is multiple-valued in $\mathbb{C}$ if $\alpha \notin \mathbb{N}_0$, we then restrict ourselves to the slit domain $S_1 := \mathbb{C} \backslash (-\infty, 0]$ and the single-valued branch of $W(z)$ in S_1 satisfying $W(1) = 1$.

For the related question of the approximation by the so-called Laurent polynomials

$$\left\{ \frac{P_{n(i)}(z)}{z^{m(i)}} \right\}_{i=0}^{\infty}, \quad \deg P_{n(i)} \leq n(i), \ (m(i), n(i) \in \mathbb{N}_0), \qquad (4.8)$$

where $\lim_{i \to \infty} \dfrac{m(i)}{n(i)} := \alpha, \alpha > 0$, we are similarly led to the question of the approximation by the weighted polynomials

$$\{z^{-\alpha n} P_n(z)\}_{n=0}^{\infty}, \deg P_n \leq n, \qquad (4.9)$$

with the only difference being in the sign in the exponent of the weight function. Thus, we can give a unified treatment of both problems by considering weighted approximation by $\{W^n(z) P_n(z)\}_{n=0}^{\infty}$, $\deg P_n \leq n$, with

$$W(z) := z^{\alpha}, \quad z \in S_1 := \mathbb{C}\setminus(-\infty, 0], \qquad (4.10)$$

where α is *any* fixed real number and where we choose, as before, the single-valued branch of $W(z)$ in S_1 satisfying $W(1) = 1$.

Theorem 4.2 [12] *Given an open set G as in (3.2(i)) with σ finite, such that $\overline{G} \subset S_1$, and given the weight function $W(z)$ of (4.10), then the pair (G, W) has the approximation property (3.3) if and only if*

$$\mu = (1 + \alpha)\omega(\infty, \cdot, \Omega) - \alpha\omega(0, \cdot, \Omega) \qquad (4.11)$$

is a positive measure, where $\omega(\infty, \cdot, \Omega)$ and $\omega(0, \cdot, \Omega)$ are, respectively, the harmonic measures with respect to the unbounded component Ω of $\overline{\mathbb{C}}\setminus \overline{G}$, at $z = \infty$ and at $z = 0$.

In some cases, when the geometric shape of G is given explicitly, we can determine the explicit form of the measure of (4.11). This is especially easy to do for disks.

Corollary 4.3 [12] *Given the disk $D_r(a) := \{z \in \mathbb{C} : |z - a| < r\}$, where $a \in (0, +\infty)$ and where $\overline{D}_r(a) \subset S_1 = \mathbb{C}\setminus(-\infty, 0]$, i.e., $r < a$, and given the weight function of (4.10), then the pair $(D_r(a), W)$ has the approximation property (3.3) if and only if*

$$r \leq r_{\max}(a, \alpha) = \begin{cases} a, & \alpha \in [-1, 0], \\ \dfrac{a}{|2\alpha + 1|}, & \alpha \in (-\infty, -1) \cup (0, \infty). \end{cases} \qquad (4.12)$$

Furthermore, if (4.12) is satisfied, then the associated measure $\mu(D_r(a), z^{\alpha})$ (see Theorem 3.1) is given by

$$d\mu(D_r(a), z^{\alpha}) = \left(1 + \alpha - \alpha \frac{a^2 - r^2}{|z|^2}\right) \frac{ds}{2\pi r}, \qquad (4.13)$$

where ds is the arclength measure on the circle $|z - a| = r$.

4.2 Jacobi and Jacobi-Type Weights

We continue along the same lines by considering weighted approximation with Jacobi weights, i.e., we set

$$W(z) := (1 - z)^{\alpha}(1 + z)^{\beta}, \quad z \in S_2 := \mathbb{C}\setminus\{(-\infty, -1] \cup [1, \infty)\}, \qquad (4.14)$$

where $\alpha, \beta \in \mathbb{R}$ are any numbers, and where we choose the branch of weight function in (4.14) such that $W(0) = 1$.

An analogue of Theorem 4.2 in this case is the following result:

Theorem 4.4 [12] *Given an open set G as in (3.2(i)) with σ finite, such that $\overline{G} \subset S_2$, and given the weight function $W(z)$ of (4.14), then the pair (G, W) has the approximation property (3.3) if and only if*

$$\mu = (1 + \alpha + \beta)\omega(\infty, \cdot, \Omega) - \alpha\omega(1, \cdot, \Omega) - \beta\omega(-1, \cdot, \Omega) \qquad (4.15)$$

is a positive measure, where Ω is the unbounded component of $\overline{\mathbb{C}} \setminus \overline{G}$.

We next state a corollary of Theorem 4.4, which deals with the explicit formula for the radius of a *largest* disk $D_r(a)$, centered at $a \in (-1, 1)$, for which $(D_r(a), W)$ has the approximation property.

Corollary 4.5 [12] *Given the disk $D_r(a) := \{z \in \mathbb{C} : |z - a| < r\}$, with $a \in (-1, 1)$ and with $\overline{D}_r(a) \subset S_2$, and given the Jacobi weight function $W(z)$ of (4.14), then the pair $(D_r(a), W)$ has the approximation property (3.3) if and only if*

$$1 + \alpha + \beta - \alpha\frac{(1-a)^2 - r^2}{|z-1|^2} - \beta\frac{(1+a)^2 - r^2}{|z+1|^2} \geq 0 \text{ on } |z - a| = r. \qquad (4.16)$$

In particular, if $\alpha \geq 0$ and $\beta \geq 0$, then the approximation property (3.3) holds if and only if

$$r \leq r_{\max}(a, \alpha, \beta) :=$$

$$\frac{\sqrt{[\alpha - \beta + a(1 + \alpha + \beta)]^2 + (1 - a^2)(1 + 2\alpha + 2\beta)} - |\alpha - \beta + a(1 + \alpha + \beta)|}{1 + 2\alpha + 2\beta}. \qquad (4.17)$$

Furthermore, if (4.16) is valid, then

$$d\mu\left(D_r(a), (1-z)^\alpha(1+z)^\beta\right)$$
$$= \left(1 + \alpha + \beta - \alpha\frac{(1-a)^2 - r^2}{|z-1|^2} - \beta\frac{(1+a)^2 - r^2}{|z+1|^2}\right)\frac{ds}{2\pi r}, \qquad (4.18)$$

where ds is the arclength measure on $|z - a| = r$.

Both weight functions introduced in (4.10) and (4.14) are special cases of the following Jacobi-type weight function:

$$W(z) := \prod_{i=1}^{p}(z - t_i)^{\alpha_i}, \qquad (4.19)$$

where $\{\alpha_i\}_{i=1}^p$ are real numbers and where $\{t_i\}_{i=1}^p \subset \mathbb{C}$ is a fixed set of distinct points. For a given open set G (as in (3.2(i))) with σ finite) such that $t_i \notin \overline{G}$, $i = 1, ..., p$, we assume that there exist p cuts, connecting each t_i with ∞. Then, we can define a single-valued branch of $W(z)$ in the p-slit complex plane which contains $\overline{G}$ in its interior. (It is not possible to specify in advance these cuts, as they necessarily depend on each preassigned open set G.)

Theorem 4.6 [12] *The pair (G, W), defined in the previous paragraph, has the approximation property (3.3) if and only if*

$$\mu = \left(1 + \sum_{i=1}^p \alpha_i\right) \omega(\infty, \cdot, \Omega) - \sum_{i=1}^p \alpha_i \omega(t_i, \cdot, \Omega) \qquad (4.20)$$

is a positive measure, where Ω is the unbounded component of $\overline{\mathbb{C}} \setminus \overline{G}$.

Furthermore, if $G = D_r(a) := \{z \in \mathbb{C} : |z - a| < r\}$ where $a \in \mathbb{C}$, then the pair $(D_r(a), W)$ has the approximation property (3.3) if and only if

$$1 + \sum_{i=1}^p \alpha_i - \sum_{i=1}^p \alpha_i \frac{|t_i - a|^2 - r^2}{|z - t_i|^2} \geq 0, \quad |z - a| = r. \qquad (4.21)$$

4.3 Exponential Weights

Let

$$W(z) := e^{-z^m}, \quad m \in \mathbb{N}. \qquad (4.22)$$

The special case $m = 1$ of the weight function (4.22) was considered in Section 2. To avoid technical complications, we shall study only the weighted approximation, with respect to the weight function $W(z) = e^{-z^m}$, in disks centered at the origin. Our next result generalizes Theorem 2.9 of Section 2.

Theorem 4.7 [12] *Given $D_r(0) := \{z \in \mathbb{C} : |z| < r\}$ and given the weight function $W(z)$ of (4.22), then the pair $(D_r(0), W)$ has the approximation property (3.3) if and only if*

$$r \leq r_{\max}(m) := (2m)^{-1/m}, \quad m \in \mathbb{N}. \qquad (4.23)$$

Moreover, if (4.23) holds, then

$$d\mu\left(D_r(0), e^{-z^m}\right) = (1 - 2mr^m \cos m\theta) \frac{d\theta}{2\pi}, \qquad (4.24)$$

where $d\theta$ is the angular measure on $|z| = r$ and where $z = re^{i\theta}$.

5 Some Proofs

Proof of Proposition 2.1 Let $z \in \overline{G} \setminus \{1\}$, where G is defined as the interior of the Szegö curve S of (1.2). Following Szegö [16], we make substitution $w = \zeta e^{1-\zeta}$ in (2.1), which gives

$$1 - e^{-nz} s_n(nz) = \frac{\sqrt{n}}{\tau_n \sqrt{2\pi}} \int_0^{ze^{1-z}} w^{n-1} \frac{\zeta(w)}{1 - \zeta(w)} dw. \qquad (5.1)$$

Our goal is to bound above the modulus of the right side of (5.1). Denoting the integral in (5.1) by I, then an integration by parts and use of the relation $w(\zeta) = \zeta e^{1-\zeta}$ give that

$$I := \int_0^{ze^{1-z}} w^{n-1} \frac{\zeta(w)dw}{1-\zeta(w)}$$

$$= \frac{w^n}{n} \frac{\zeta(w)}{1-\zeta(w)} \bigg|_0^{ze^{1-z}} - \int_0^{ze^{1-z}} \frac{w^n}{n} \frac{\zeta'(w)dw}{(1-\zeta(w))^2}$$

$$= \frac{(ze^{1-z})^n}{n} \frac{z}{1-z} - \frac{1}{n} \int_0^z \frac{(\zeta e^{1-\zeta})^n}{(1-\zeta)^2} d\zeta$$

$$= \frac{1}{n(1-z)} \left(z(ze^{1-z})^n + (z-1) \int_0^z \frac{(\zeta e^{1-\zeta})^n}{(1-\zeta)^2} d\zeta \right).$$

To bound above $|I|$, assume that $z = (x + iy) \in \overline{G}\setminus\{1\}$, and choose the path of integration in the integral of I to consist of the two intervals $[0, x]$ and $[x, x + iy]$. Then, as $|z| \leq 1$ and $|ze^{1-z}| \leq 1$ for all points of $\overline{G}$,

$$|I| \leq \frac{1}{n|1-z|} \left\{ 1 + |z-1| \left(\int_0^{|x|} \frac{dt}{(1-t)^2} + \int_0^{|y|} \frac{ds}{(1-|x|)^2} \right) \right\}$$

$$= \frac{1}{n|1-z|} \left\{ 1 + |z-1| \left(\frac{|x|}{1-|x|} + \frac{|y|}{(1-|x|)^2} \right) \right\}$$

$$\leq \frac{1}{n|1-z|} \left\{ 1 + |z-1| \left(\frac{1}{1-|x|} + \frac{|y|}{(1-|x|)^2} \right) \right\}.$$

Then, it can be verified that the square, with vertices ± 1 and $\pm i$, contains $\overline{G}$. (This is shown in Fig. 4 below.) This geometrically implies that

$$|y| \leq 1 - |x| \text{ and } 1 - |x| \leq |1-z| \leq \sqrt{2}(1-|x|).$$

Inserting these inequalities into the upper bound above for $|I|$ yields

$$|I| \leq \frac{1}{n|1-z|} \left\{ 1 + 2\frac{|z-1|}{1-|x|} \right\} \leq \frac{1}{n|1-z|} \{1 + 2\sqrt{2}\} \leq \frac{4}{n|1-z|},$$

for any $z = x + iy \in \overline{G}$. This bound, applied to (5.1), then gives the desired result of (2.4) of Proposition 2.1. □

Proof of Theorem 2.3 First, fix a small δ with $1 > \delta > 0$ and consider the domain $G_\delta := G\setminus\{z : |z-1| \leq \delta\}$. It follows immediately from (2.4) that

$$\|e^{-nz}s_n(nz) - 1\|_{\overline{G_\delta}} \leq \frac{4}{\sqrt{2\pi n}} \left\|\frac{1}{z-1}\right\|_{\overline{G_\delta}} = \frac{4}{\sqrt{2\pi n\delta}} \quad (n \in \mathbb{N}),$$

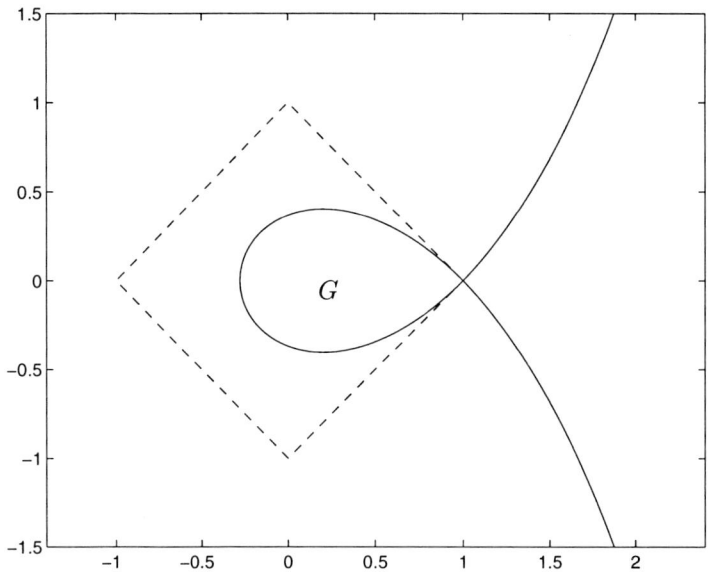

FIG. 4. The Szegő domain G and the covering square

so that
$$\|e^{-nz}s_n(nz) - 1\|_{\overline{G}_\delta} \to 0, \text{ as } n \to \infty. \tag{5.2}$$
Multiplying (2.4) by e^{-kz}, we similarly observe that, for *any* fixed $k = 0, 1, 2, \ldots$,
$$\|e^{-(n+k)z}s_n(nz) - e^{-kz}\|_{\overline{G}_\delta} \leq \frac{4}{\sqrt{2\pi n\delta}}\|e^{-kz}\|_{\overline{G}_\delta} \quad (n \in \mathbb{N}),$$
so that
$$\|e^{-(n+k)z}s_n(nz) - e^{-kz}\|_{\overline{G}_\delta} \to 0, \text{ as } n \to \infty. \tag{5.3}$$
This means that e^{-kz}, for any $k \geq 0$, can be uniformly approximated on $\overline{G}_\delta$ by the weighted polynomials $\{e^{-nz}s_{n-k}((n-k)z)\}_{n=k}^\infty$. Therefore, any complex polynomial in e^{-z}, say $Q_m(e^{-z}) = \sum_{j=0}^m c_j e^{-jz}$, can also be uniformly approximated on $\overline{G}_\delta$ by such weighted polynomials $e^{-nz}P_n(z)$. In fact, it is easy to see from (5.3) that these polynomials $P_n(z)$, in this case, can be chosen to be
$$P_n(z) := \sum_{j=0}^m c_j s_{n-j}((n-j)z), \quad n \geq m.$$

If we show that any function $f(z)$, which is analytic in G and continuous on compact subsets of $\overline{G}\setminus\{1\}$, is uniformly approximable on $\overline{G}_\delta$ by polynomials

$Q_m(e^{-z})$, then (2.6) of Theorem 2.3 will follow. Indeed, as ∂G_δ is a Jordan curve (with interior G_δ), so is its image, ∂H_δ (with interior H_δ) in the t-plane, under the conformal mapping $\Phi(z) := e^{-z} = t$. As can be readily verified, for any δ with $0 \leq \delta \leq 1$, the image of $\overline{G}_\delta$, under $\Phi(z) = t$, lies in the open right-half plane of the t-plane, and is symmetric about the positive real axis. Hence, on cutting the t-plane along the negative real axis, then $f(-\log t)$ is analytic and single-valued in H_δ and continuous on ∂H_δ. Thus, by Mergelyan's Theorem (cf. Gaier [3, p. 97]), $f(-\log t)$ can be uniformly approximated on $\overline{H}_\delta$ by the polynomials $Q_m(t)$. But this means that $f(z)$ can be uniformly approximated by $Q_m(e^{-z})$ on $\overline{G}_\delta$.

To prove the second assertion of Theorem 2.3, we note, on multiplying (2.4) by $(z-1)e^{-kz}$, that

$$\|e^{-(n+k)z}((z-1)s_n(nz)) - (z-1)e^{-kz}\|_{\overline{G}} \leq \frac{4\|e^{-kz}\|_{\overline{G}}}{\sqrt{2\pi n}},$$

which implies that any function of the form $(z-1)e^{-kz}$, $k \geq 1$, can be approximated by weighted polynomials $e^{-nz}P_n(z)$, uniformly on $\overline{G}$. It follows, for any polynomial $Q_m(t)$ with $Q_m(0) = 0$, that $(z-1)Q_m(e^{-z})$ is uniformly approximable by weighted polynomials $e^{-nz}P_n(z)$ on $\overline{G}$. Now, assume in addition that $f(z)$ is analytic at $z=1$. On defining the function

$$v(z) := \frac{f(z)e^z}{z-1},$$

it follows, since $f(1) = 0$ by hypothesis, that $v(z)$ is analytic in G and continuous on $\overline{G}$. Then, by the previous argument (with G_δ and H_δ being replaced, respectively by G and H) and with the same mapping $\Phi(z) := e^{-z} = t$, it similarly follows that $v(-\log t)$ can be uniformly approximated on $\overline{H}$ by the polynomials $\tilde{Q}_{m-1}(t)$, so that

$$\left\|\frac{f(z)e^z}{z-1} - \tilde{Q}_{m-1}(e^{-z})\right\|_{\overline{G}} = \|v(-\log t) - \tilde{Q}_{m-1}(t)\|_{\overline{H}} \to 0,$$

as $m \to \infty$. On setting $Q_m(t) := t\tilde{Q}_{m-1}(t)$ so that $Q_m(0) = 0$, the above display, after multiplying through by $(z-1)e^{-z}$, gives that

$$\|f(z) - (z-1)Q_m(e^{-z})\|_{\overline{G}} \to 0 \text{ as } m \to \infty,$$

which shows that $f(z)$ is uniformly approximable on $\overline{G}$ by weighted polynomials $e^{-nz}P_n(z)$. To complete the proof, we now drop the hypothesis that $f(z)$ is analytic at $z = 1$. Let $P_n(z)$ be the best uniform approximation from π_n to $f(z)$ on $\overline{G}$. By Mergelyan's Theorem again,

$$\lim_{n \to \infty} \|f(z) - P_n(z)\|_{\overline{G}} = 0.$$

For each $n \geq 0$, define $\tilde{P}_n(z) := P_n(z) - P_n(1)$, so that $\tilde{P}_n(1) = 0$. Because $f(1) = 0$, we see that

$$|P_n(1)| = |f(1) - P_n(1)| \leq \|f - P_n\|_{\overline{G}},$$

which implies that

$$\|f - \tilde{P}_n\|_{\overline{G}} \leq \|f - P_n\|_{\overline{G}} + |P_n(1)| \leq 2\|f - P_n\|_{\overline{G}},$$

i.e., $f(z)$ can be uniformly approximated in $\overline{G}$ by $\tilde{P}_n(z)$. But as our previous proof can be applied to each $\tilde{P}_n(z)$, it follows that $f(z)$ can be uniformly approximated on $\overline{G}$ by the weighted polynomials. □

Proof of Theorem 2.5 Given the pair of numbers (r, R) with $0 < r < R \leq 1$, suppose that $f(z)$ is analytic in G_R. For each $n \geq 0$, let $\left\{z_k^{(n+1)}\right\}_{k=1}^{n+1}$ be $n+1$ points (to be specified below) such that $\left\{z_k^{(n+1)}\right\}_{k=1}^{n+1} \subset G_R$. Then, from the Hermite interpolation formula, the polynomial $P_n(z)$, which interpolates $e^{nz}f(z)$ in the $n+1$ points $\left\{z_k^{(n+1)}\right\}_{k=1}^{n+1}$, is given (cf. [19, p. 50]) by

$$e^{nz}f(z) - P_n(z) = \frac{\omega_{n+1}(z)}{2\pi i} \int_{S_{R-\epsilon}} \frac{f(t)e^{nt}dt}{(t-z)\omega_{n+1}(t)}, \tag{5.4}$$

where $\omega_{n+1}(z) := \prod_{k=1}^{n+1}\left(z - z_k^{(n+1)}\right)$ and where $z \in G_{R-\epsilon}$; here, $\epsilon > 0$ is chosen sufficiently small so that $\left\{z_k^{(n+1)}\right\}_{k=1}^{n+1} \subset G_{R-\epsilon}$. Dividing by e^{nz} in (5.4) gives

$$f(z) - e^{-nz}P_n(z) = \frac{e^{-nz}\omega_{n+1}(z)}{2\pi i} \int_{S_{R-\epsilon}} \frac{f(t)dt}{(t-z)e^{-nt}\omega_{n+1}(t)}, \tag{5.5}$$

for $z \in G_{R-\epsilon}$.

Let $\nu_n(\omega_n)$ be the normalized counting measure of the zeros of $\omega_n(z)$, i.e.,

$$\nu_n(\omega_n) = \frac{1}{n}\sum_{k=1}^{n}\delta_{z_k^{(n)}} \quad (n \in \mathbb{N}), \tag{5.6}$$

where δ_z is the unit point mass at z and where all zeros are counted according to their multiplicities. Then, from the definition in (3.4),

$$|\omega_n(z)| = \exp\left\{-nU^{\nu_n(\omega_n)}(z)\right\} \quad (n \in \mathbb{N}). \tag{5.7}$$

For each r with $0 < r < R$, we now choose an interpolation scheme in (5.4) which satisfies

$$\left\{z_k^{(n)}\right\}_{k=1}^{n} \subset S_r \quad (n \in \mathbb{N}), \tag{5.8}$$

and for which
$$\nu_n(\omega_n) \stackrel{*}{\to} \omega(0,\cdot,G_r), \quad \text{as } n \to \infty. \tag{5.9}$$

(Here, $\omega(0,\cdot,G_r)$ is the harmonic measure; see [10] and [18]. In addition, the convergence in (5.9) of the discrete measures is in terms of weak* convergence of measures, i.e., a sequence of Borel measures $\{\mu_n\}_{n\in\mathbb{N}}$ on $\mathbb{C}$ converges to a measure μ, as $n \to \infty$, in the *weak* topology* (written $\mu_n \stackrel{*}{\to} \mu$) if

$$\lim_{n\to\infty} \int f \, d\mu_n = \int f \, d\mu$$

for any continuous function f on $\mathbb{C}$ having compact support.) As an example of an interpolation where (5.8) and (5.9) are valid, one can take the preimages of equally spaced points on $|w| = r$ under the conformal map $w = \varphi(z) = ze^{1-z}$, i.e., for $\psi := \varphi^{[-1]}$, we define

$$z_k^{(n)} := \psi\left(re^{i\frac{2\pi k}{n}}\right) \quad (1 \le k \le n, \quad n \in \mathbb{N}). \tag{5.10}$$

It follows from (5.7)–(5.9) that

$$\lim_{n\to\infty} |\omega_n(z)|^{1/n} = \lim_{n\to\infty} \exp\left\{-U^{\nu_n(\omega_n)}(z)\right\} \\ = \exp\left\{-U^{\omega(0,\cdot,G_r)}(z)\right\}, \tag{5.11}$$

which holds locally uniformly in $\mathbb{C}\setminus\overline{G}_r$. Taking any ϵ small enough so that $r + \epsilon < R - \epsilon$, we estimate the difference in (5.5) by

$$\|f(z) - e^{-nz}P_n(z)\|_{\overline{G}_r} \le \|f(z) - e^{-nz}P_n(z)\|_{\overline{G}_{r+\epsilon}}$$

$$\le \frac{\|e^{-nz}\omega_{n+1}(z)\|_{S_{r+\epsilon}} \|f\|_{S_{R-\epsilon}}}{2\pi \text{dist}(S_{r+\epsilon}, S_{R-\epsilon}) \cdot \min_{t \in S_{R-\epsilon}} |e^{-nt}\omega_{n+1}(t)|}.$$

Thus, we obtain, by (5.11), that (for details, see Pritsker and Varga [11])

$$\limsup_{n\to\infty} \|f(z) - e^{-nz}P_n(z)\|_{S_r}^{1/n} \le \frac{e^{\log(r+\epsilon)-1}}{e^{\log(R-\epsilon)-1}} = \frac{r+\epsilon}{R-\epsilon}.$$

Letting $\epsilon \to 0$, this gives (2.12) of Theorem 2.5.

To show that the converse part of Theorem 2.5 is valid, suppose that (2.12) holds true for r with $0 < r < R \le 1$. Then, the rest of the proof is a classical converse theorem argument (see [19, p. 81], for example). By the uniform convergence on $\overline{G}_r$, the function $f(z)$ can be represented, in a telescopic series, as

$$f(z) = e^{-nz}P_n(z) + \sum_{k=n}^{\infty} \left(e^{-(k+1)z}P_{k+1}(z) - e^{-kz}P_k(z)\right), z \in \overline{G}_r. \tag{5.12}$$

Thus,
$$|f(z) - e^{-nz}P_n(z)| \leq \sum_{k=n}^{\infty} \left|e^{-(k+1)z}P_{k+1}(z) - e^{-kz}P_k(z)\right|. \tag{5.13}$$

For any $\epsilon > 0$, we have from (2.12) that
$$\|f(z) - e^{-kz}P_k(z)\|_{\overline{G}_r} \leq \left(\frac{r}{R-\epsilon}\right)^k,$$

if $k \geq n$ is sufficiently large. This gives that
$$\|e^{-(k+1)z}P_{k+1}(z) - e^{-kz}P_k(z)\|_{\overline{G}_r}$$
$$\leq \|f(z) - e^{-kz}P_k(z)\|_{\overline{G}_r} + \|f(z) - e^{-(k+1)z}P_{k+1}(z)\|_{\overline{G}_r}$$
$$\leq C_1 \left(\frac{r}{R-\epsilon}\right)^k, \quad k \geq n,$$

where C_1 is a constant, independent of k. Using the result of [11, Corollary 4.2], it is known, for any polynomial $P_n(z)$ with $\deg P_n \leq n$, that
$$|e^{-nz}P_n(z)| \leq \|e^{-nz}P_n(z)\|_{S_r} \cdot \left(\frac{|\varphi(z)|}{r}\right)^n, \tag{5.14}$$

for any $z \in \mathbb{C} \setminus G_r, n \in \mathbb{N}_0$, and $0 < r \leq 1$. Applying this to the previous display gives
$$|e^{-(k+1)z}P_{k+1}(z) - e^{-kz}P_k(z)| \leq C_1 \left(\frac{r}{R-\epsilon}\right)^k \left(\frac{|\varphi(z)|}{r}\right)^k$$
$$= C_1 \left(\frac{|\varphi(z)|}{R-\epsilon}\right)^k, \quad k \geq n.$$

If $|\varphi(z)| = R - 2\epsilon$, i.e., $z \in S_{R-2\epsilon}$, then the telescopic series (5.12) converges to the analytic continuation of $f(z)$ in $G_{R-2\epsilon}$. Thus, from (5.13) and the above inequalities,
$$\|f(z) - e^{-nz}P_n(z)\|_{\overline{G}_{R-2\epsilon}} \leq C_2 \left(\frac{R-2\epsilon}{R-\epsilon}\right)^n, \tag{5.15}$$

for any sufficiently large n. Hence, the sequence $\{e^{-nz}P_n(z)\}_{n=0}^{\infty}$ converges to the analytic continuation of $f(z)$, uniformly on $\overline{G}_{R-2\epsilon}$. Since $\epsilon > 0$ can be taken arbitrarily small, then $f(z)$ must be analytic in G_R. □

Proof of Corollary 2.6 If $f(z)$ is analytic in G_R, then by Theorem 2.5,
$$\limsup_{n \to \infty} \left[E_n^{\exp(-z)}(f, \overline{G}_r)\right]^{1/n} \leq \frac{r}{R}, \tag{5.16}$$

where $0 < r < R < 1$. However, strict inequality in (5.16) is equivalent to the analyticity of $f(z)$ in G_ρ, for some ρ with $R < \rho < 1$, by virtue of Theorem 2.5. Thus, $f(z)$ has a singularity on S_R if and only if equality holds in (5.16). □

Proof of Theorem 2.7 Suppose that $f(z)$ is analytic in the interior of $\zeta + E$ and continuous on $\{\zeta + E\}$, where E is a compact set. Then, $g(t) := f(t + \zeta)$ is analytic in the interior of E and continuous on E, which implies by hypothesis that $g(t)$ can be uniformly approximated on E by $\{e^{-nt} P_n(t)\}_{n=0}^\infty$. Thus, with $z = t + \zeta$, $f(z)$ can be approximated on $\{\zeta + E\}$ by the weighted polynomials

$$e^{-n(z-\zeta)} P_n(z - \zeta) = e^{-nz} \left(e^{n\zeta} P_n(z - \zeta)\right) \quad (n \in \mathbb{N}_0). \quad \square$$

Proof of Theorem 2.8 Because of Theorem 2.7, we may assume that the domain H is such that $\overline{G} \subset H$. Further, assume to the contrary, that, for some $f(z) \not\equiv 0$ which is analytic in H, there exists a sequence of polynomials $\{P_n(z)\}_{n=0}^\infty$, $\deg P_n \le n$, such that

$$\lim_{n \to \infty} \left\| f(z) - e^{-nz} P_n(z) \right\|_{\overline{G}} = 0. \tag{5.17}$$

It follows that

$$\lim_{n \to \infty} \left\| e^{-nz} P_n(z) \right\|_{\overline{G}} = \|f\|_{\overline{G}} \ne 0. \tag{5.18}$$

But (5.14), for the case $r = 1$, and (5.18) immediately give that

$$\lim_{n \to \infty} \left| e^{-nz} P_n(z) \right| = 0, \quad \text{for any } z \in \Omega_0, \tag{5.19}$$

where the convergence in (5.19) is locally uniform in Ω_0. Thus, the convergence of $\{e^{-nz} P_n(z)\}_{n=0}^\infty$ to $f(z) \not\equiv 0$, locally uniformly in H, is impossible because $H \cap \Omega_0 \ne \emptyset$. □

Proof of the Second Remark after Theorem 3.1 To prove the second statement in this remark, suppose then that $W(z_0) = 0$ with $z_0 \in G_\ell$, where $W(z) \not\equiv 0$ in G_ℓ, and suppose, given an analytic function $f(z)$ in G, that polynomials $\{P_n(z)\}_{n=0}^\infty$ can be found such that $\{W^n(z) P_n(z)\}_{n=0}^\infty$ converges to $f(z)$, locally uniformly in G. As $W(z) \not\equiv 0$, we can choose $R > 0$ such that $D_R(z_0) := \{z \in \mathbb{C} : |z - z_0| < R\}$ satisfies $\overline{D}_R(z_0) \subset G_\ell$ and that

$$M := \min_{|z - z_0| = R} |W(z)| > 0. \tag{5.20}$$

Then, by the locally uniform convergence of $\{W^n(z) P_n(z)\}_{n=0}^\infty$ to $f(z)$,

$$\|P_n\|_{\overline{D}_R(z_0)} = \|P_n\|_{\partial D_R(z_0)} \le \|W^n P_n\|_{\partial D_R(z_0)} \|W^{-n}\|_{\partial D_R(z_0)} \le \frac{\|f\|_{\partial D_R(z_0)} + 1}{M^n}, \tag{5.21}$$

for all $n \in \mathbb{N}$ sufficiently large. Since $W(z_0) = 0$, we can find an $r \in (0, R)$ such that

$$m := \|W\|_{\overline{D}_r(z_0)} < M. \tag{5.22}$$

Using (5.21) and (5.22), we obtain

$$\|W^n P_n\|_{\overline{D}_r(z_0)} \leq \|W\|^n_{\overline{D}_r(z_0)} \|P_n\|_{\overline{D}_r(z_0)} \leq \left(\frac{m}{M}\right)^n (\|f\|_{\partial D_R(z_0)} + 1) \to 0,$$

as $n \to \infty$.

But because of the locally uniform approximation of $f(z)$ by $\{W^n(z)P_n(z)\}_{n=0}^{\infty}$, it follows that $f(z) \equiv 0$ for any $z \in \overline{D}_r(z_0)$, which implies, by the uniqueness theorem, that $f(z) \equiv 0$ in G_ℓ. □

6 Further Remarks and Open Problems

Theorem 3.1 gives a rather complete answer to the question on weighted approximation by $W^n(z)P_n(z)$ in open sets of the complex plane. It is then very natural to consider the uniform approximation by such weighted polynomials on *compact* sets, aiming at an analogue (generalization) of Mergelyan's Theorem (see [19, p. 367]). Let $E \subset \mathbb{C}$ be a compact set with connected complement $\overline{\mathbb{C}} \backslash E$. We denote the set of all functions, analytic interior to E and continuous on E, by $A(E)$. Let $W \in A(E)$, with $W(z) \neq 0$ for any $z \in E$.

Problem *Give a necessary and sufficient condition for the pair (E, W) to have the following approximation property:*

For any $f \in A(E)$, there exist polynomials $\{P_n(z)\}_{n=0}^{\infty}$, with $\deg P_n \leq n$, such that

$$\lim_{n \to \infty} \|f - W^n P_n\|_E = 0. \tag{6.1}$$

Obviously, the classical uniform approximation by polynomials (Mergelyan's Theorem) corresponds to $W(z) \equiv 1$, $z \in E$. We observe that (3.5) of Theorem 3.1, holding with $G = \text{Int} E$, is a necessary condition for (6.1). Let us also remark that this problem is open even in the case when E is a subset of the real line, such as an interval (see [13, 17] for background and general results).

An even more general approach is to consider the approximation problem in (6.1) with polynomials replaced by rational functions. Certain results concerning such weighted rational approximation have been obtained by Borwein and Chen [1] in the complex plane.

Bibliography

1. Borwein, P. B. and Chen, W. (1995). Incomplete rational approximation in the complex plane. *Constr. Approx.*, **11**, 85–106.
2. Carpenter, A. J., Varga, R. S. and Waldvogel, J. (1991). Asymptotics for the zeros of e^z. I. *Rocky Mount. J. of Math.*, **21**, 99–119.
3. Gaier, D. (1987). *Lectures on Complex Approximation*. Birkhäuser.

4. Golitschek, M. v. (1980). Approximation by incomplete polynomials. *J. Approx. Theory*, **28**, 155–160.
5. Henrici, P. (1977). *Applied and Computational Complex Analysis*, vol. 2. Wiley.
6. R. Jentzsch, R. (1917). Untersuchungen zur Theorie der Folgen analytischer Funktionen. *Acta Math.*. **41**, 219–251.
7. Landkof, N. S. (1972). *Foundations of Modern Potential Theory*. Springer.
8. Lorentz, G. G. (1977). Approximation by incomplete polynomials (problems and results). In: *Padé and Rational Approximations: Theory and Applications* (ed. E. B. Saff and R. S. Varga), pp. 289–302. Academic Press.
9. Marden, M. (1966). *Geometry of Polynomials*, Mathematical Surveys No. 3. Amer. Math. Soc., Providence.
10. Nevanlinna, R. (1970). *Analytic Functions*. Springer.
11. Pritsker, I. E. and Varga, R. S. (1997). The Szegő curve, zero distribution and weighted approximation. *Trans. Amer. Math. Soc.*. **349**, 4085–4105.
12. Pritsker, I. E. and Varga, R. S. (1997). Weighted polynomial approximation in the complex plane. *Constr. Approx.*, to appear.
13. Saff, E. B. and Totik, V. (1997). *Logarithmic Potentials with External Fields*. Springer.
14. Saff, E. B. and Varga, R. S. (1978). On incomplete polynomials. In: *Numerische Methoden der Approximationstheorie*. (ed. L. Collatz, G. Meinardus, and H. Werner), ISNM 42, pp. 281–298. Birkhäuser.
15. G. Szegő, G. (1922). Über die Nullstellen von Polynomen die in einem Kreise gleichmässig konvergieren. *Sitzungsber. Berl. Math. Ges.*. **21**, 59–64.
16. Szegő, G. (1924). Über eine Eigenschaft der Exponentialreihe. *Sitzungsber. Berl. Math. Ges.*. **23**, 50–64.
17. Totik, V. (1994). *Weighted Approximation with Varying Weights*. Lecture Notes in Math., vol. 1569. Springer.
18. Tsuji, M. (1959). *Potential Theory in Modern Function Theory*. Maruzen, Tokyo.
19. Walsh, J. L. (1969). *Interpolation and Approximation by Rational Functions in the Complex Domain*. Colloquium Publications, vol. 20. Amer. Math. Soc., Providence.

9
Existence and the Singular Relaxation Limit for the Inviscid Hydrodynamic Energy Model

Gui-Qiang Chen and Joseph W. Jerome
Northwestern University

and

Bo Zhang
Stanford University

Abstract

It is well known to particle physicists that scattering-dominated transport processes lead to drift-diffusion regimes. In this chapter, we give a mathematical study of this result, beginning with the full hydrodynamic model, and identifying scaled drift-diffusion as the zero relaxation limit. In order to provide a more complete background of drift-diffusion, we describe steady-state parameter regions in the introduction. This material is not essential to the chapter.

1 Introduction

In this chapter, we shall study a hydrodynamic model for charged carrier transport, and its singular relaxation drift-diffusion limit. Although the emphasis of the chapter will be on the theory of the hydrodynamic model and its limit, and although the theory will proceed for the evolution system, we shall present in the introduction an idealized steady-state drift-diffusion model, so that the reader unfamiliar with this area can gain some appreciation for the rich and diverse behavior possible even in this special case. After presenting this model, we shall then discuss, in the remaining part of the introduction, the context in which the principal results of the chapter are derived. Before proceeding to the steady-state model, we shall describe the setting of charge transport in the principal application areas of semiconductors and ionic channels.

It is helpful to view semiclassical semiconductor modeling as a hierarchical structure, in which the Boltzmann transport equation forms the summit, and the drift-diffusion model the base; intermediate are systems derived from moments of the Boltzmann transport equation. The hydrodynamic model is an example, and it offers the promise of simulation detail, as well as feasibility. It does have the drawback of requiring adequate closure assumptions and accurate representations for the average collision mechanisms. In addition, the model has hyperbolic as well as parabolic components for the moment equations, in the transient case, and thus issues of shock capturing mechanisms arise. On the other hand, the drift-diffusion model has only parabolic components, exclusive of the Poisson equation, which must be adjoined to any model in the hierarchy. An interesting derivation of the drift-diffusion model from the Boltzmann transport equation, making use of the diffusion approximation, is carried out in [10]. This model is by far the best understood of the models. It was introduced by van Roosbroeck in 1950 [28] as a conservation system for electron and hole carriers, with the ambient electric field determined from the Poisson equation. This model is closely related to the older (by sixty years) ionic transport model of Nernst and Planck, which is described in detail in [24]. It is of interest that the current–voltage curves in biological membrane channels are derived from mechanisms closely related to those described here, with similar systems of equations involved in the modeling. The reader is invited to consult [23] for scientific detail.

1.1 Idealized Drift-Diffusion

Physically, this is a case of constant temperature in which friction, caused by collisions, dominates. We select the case of *vanishing* permanent charge and *simple* boundary conditions. It describes:

- electrons and holes in a pure crystalline lattice; or
- positive and negative ions in a vacuum environment,

together with equilibrium boundary conditions. The steady system is defined by continuity equations for

(1) electric potential ϕ; charge flux: $-(\text{dielectric}) \times \phi'$;

(2) negative carrier (electron/anion) concentration n;

$$\text{charge: } -\bar{e}, \quad \text{current: } J_n = \bar{e}D_n(\underbrace{n'}_{\text{diffusion}} \underbrace{-n\phi'}_{\text{drift}});$$

(3) positive carrier (hole/cation) concentration p;

$$\text{charge: } \bar{e}, \quad \text{current: } J_p = \bar{e}D_p(p' + p\phi').$$

The equations (with vanishing permanent charge) are given by:

$$\epsilon^2 \phi'' - n + p = 0, \tag{1.1}$$
$$J_n = \text{const.}, \tag{1.2}$$
$$J_p = \text{const.}, \tag{1.3}$$

in $(0,1)$, with boundary conditions

$$p(0) = p_L > 0, \quad n(0) = n_L > 0, \tag{1.4}$$
$$p(1) = p_R > 0, \quad n(1) = n_R > 0, \tag{1.5}$$
$$\phi(0) = \phi_0, \quad \phi(1) = \phi_1. \tag{1.6}$$

Without loss of generality, take $\phi_1 = 0$. Above, D_n and D_p denote diffusion coefficients. We highlight here the following definitions:

- potential difference: $V = \phi_0$;
- physical current: $I = J_n + J_p$;
- negative of (mass) flux: $J = J_n - J_p$;
- ϵ is the quotient of two lengths, intrinsic and extrinsic.

A notation in terms of dimensionless ratios is given by:
$$\rho_L = \frac{p_L}{n_L}, \quad \rho_R = \frac{p_R}{n_R}. \tag{1.7}$$

System (1.1)–(1.6) has *simple* boundary values if $\rho_L = \rho_R = 1$. In this case, denote the common values by c_L, c_R, respectively.

There are four modes for simple boundary conditions, as is demonstrated in [22]:

Case 1. $c_L \leq c_R$, $V \geq 0$.

(1) $IV \geq 0$, $J \geq 0$; (2) $\phi_x \leq 0$; (3) $n \geq p$; (4) $n_x, p_x \geq 0$.

Case 2. $c_L \leq c_R$, $V \leq 0$.

(1) $IV \geq 0$, $J \geq 0$; (2) $\phi_x \geq 0$; (3) $n \leq p$; (4) $n_x, p_x \geq 0$.

Case 3. $c_L \geq c_R$, $V \leq 0$.

(1) $IV \geq 0$, $J \leq 0$; (2) $\phi_x \geq 0$; (3) $n \geq p$; (4) $n_x, p_x \leq 0$.

Case 4. $c_L \geq c_R$, $V \geq 0$.

(1) $IV \geq 0$, $J \leq 0$; (2) $\phi_x \leq 0$; (3) $n \leq p$; (4) $n_x, p_x \leq 0$.

Figure 1 illustrates a typical case. In all cases, the solution is unique, as demonstrated in [22].

1.2 The Hydrodynamic Model

This model may alternatively be defined from a microscopic starting point, the semiclassical Boltzmann equation, in which moments are computed with respect to group velocity, and the system closed at three equations with constitutive relations among the averaged quantities, or it can be derived from a macroscopic starting point, in which the system then becomes a conservation law system for a charged fluid, incorporating particle number, momentum, and energy balance equations. Both forms of derivation are given by Jerome [14, Chap. 2]. The result is the following system, with a copy for each charge species:

$$\begin{cases} \partial_t \rho + \nabla \cdot (\rho v) = C_\rho, \\ \partial_t p + v(\nabla \cdot p) + (p \cdot \nabla) v = -\frac{\bar{e}}{\bar{m}} \rho F - \nabla(\rho k_b T / \bar{m}) + C_p, \\ \partial_t E + \nabla \cdot (v\, E) = -\frac{\bar{e}}{\bar{m}} \rho v \cdot F - \nabla \cdot (v \rho k_b T / \bar{m}) + \nabla \cdot (\kappa \nabla T) + C_E. \end{cases} \tag{1.8}$$

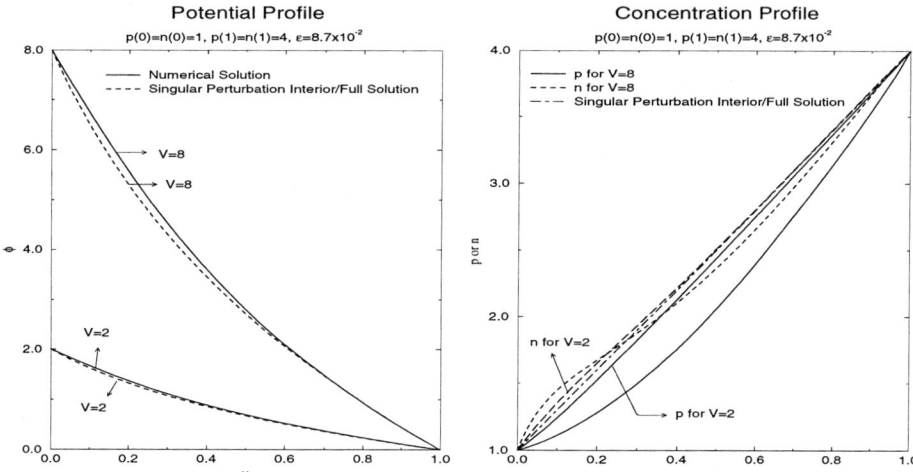

FIG. 1. The electric potential profile and concentration profiles for the simple boundary condition, Case 1. Dimensionless concentration and potential units employed. Reproduced from [2]

Here, ρ denotes particle mass density, related to concentration n and effective mass $\bar{m}$ via $\rho = \bar{m}n$, p denotes particle momentum density, related to velocity v through $p = \rho v$, and E the mechanical energy density. F denotes the electric field, T the carrier temperature, κ the heat conductivity, k_b Boltzmann's constant, and C_ρ, C_p, and C_E denote relaxation expressions. The systems are coupled through the Poisson electrostatic equation as well. The momentum relaxation time τ_p is given via

$$\frac{p}{\tau_p} := -C_p.$$

The energy relaxation time τ_w is given via

$$-\frac{W - W_0}{\tau_w} := C_W.$$

Here, W_0 denotes the rest energy, $\frac{3}{2}nk_b\bar{T}$, where $\bar{T}$ is the lattice temperature. Note that the relaxation time reciprocals appearing here have the units of frequency, and suggest time scales over which collisions are likely to occur in influencing momentum or energy. The forms for the relaxation times used in [1] on the basis of higher-order moments are:

$$\tau_p = c_p/T,$$
$$\tau_w = c_w \frac{T}{T + \bar{T}} + \frac{1}{2}\tau_p.$$

Here, c_p and c_w are physical constants. The hydrodynamic model was first derived by Bløtekjær [3] for semiconductor devices. It can describe multiple species of charge carriers (e.g., upper and lower valley electrons and holes) by adding extra copies of (1.8). This model also governs an important biological process in which ions move into biological cells through pores in proteins that are holes in dielectrics [5]. The model (1.8) plays an important role in simulating the behavior of charged carriers in submicron semiconductor devices; some of its computational and physical aspects have been discussed in [11, 25, 16]. The existence of solutions and the convergence of Newton's method to the steady-state subsonic electron flow have been proved in [13]. Also, an electron shock wave in the case of steady state for (1.8) for submicron semiconductor devices was first simulated in [12].

1.3 Regimes Defined by Damping

We identify three essential regimes defined by the hydrodynamic model:

(1) the full hydrodynamic regime;
(2) a regime in which inertial effects are negligible over the momentum relaxation time scale;
(3) a regime, contained within the preceding, in which temperature variation is negligible, so that heat flux is insignificant. This corresponds to the drift-diffusion regime. It is characterized as friction-dominated.

In the first regime, friction is not dominant. In the second, resistance to momentum transitions is significant over the relaxation time scale. In the third regime, additional resistance to heat flow (temperature gradients) is significant. Regime (2) may be defined via a critical limit of the hydrodynamic model, as we now explain. The first and third equations in (1.8) remain as in the hydrodynamic model. However, the momentum equation is rewritten, and the limit,

$$\tau_p \to 0,$$

is taken to obtain a constitutive relation for the current, given by:

$$J = \frac{\bar{e}D}{\bar{m}} \left[\frac{T}{\bar{T}} \nabla \rho + \rho \nabla \left(\frac{T}{\bar{T}} - \frac{\bar{e}}{k_b \bar{T}} \phi \right) \right]. \qquad (1.9)$$

Here, D is the diffusion coefficient and ϕ is the electrostatic potential. This is the interpretation of the inertial approximation. The third regime follows from the second when isothermal conditions prevail.

These are strictly heuristic considerations. It is the purpose of this chapter to derive the third regime in a completely rigorous fashion, as a scaled limit of the one-carrier model as τ_p and τ_w tend to zero in a controlled manner. In fact, the fundamental inequality relating τ_p and τ_w, and required by the theory developed below, is satisfied by the representations given above.

We close the introduction by exhibiting in Figures 2 and 3 the results of a series of simulations carried out by decreasing τ_w via increasing a parameter, the saturation velocity, v_s, which is related to c_w in an inverse square fashion. The application is to the biological regime, that of ionic channels. The tendency of regime (1) toward regime (3) is evident in these pictures.

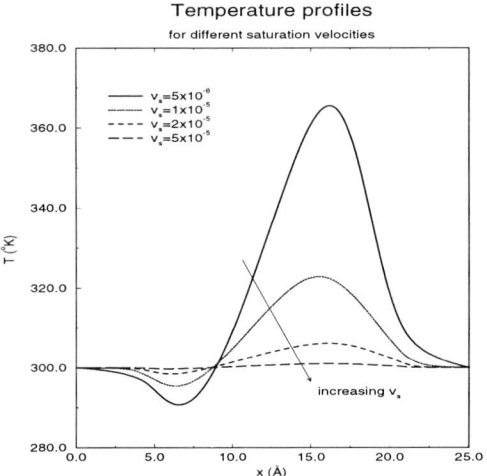

FIG. 2. The solid curve gives the temperature for $v_s = 5 \times 10^{-6}$, the dotted curve is for $v_s = 10^{-5}$, the short-dashed curve is for $v_s = 2 \times 10^{-5}$, and the long-dashed curve is for $v_s = 5 \times 10^{-5}$. The decrease of temperature coincides with efficient damping of energy exchange. The units of v_s are $\mu m/ps$. From [5]

2 Energy Method for a Scalar Equation with Damping

In this section, through a scalar model with damping, we present an efficient method, the energy method, for determining the global existence, uniqueness, and asymptotic behavior of classical smooth solutions for nonlinear partial differential equations with certain dissipation. Such a method has been successfully applied to problems of classical smooth solutions for nonlinear wave equations in [17], nonlinear hyperbolic Volterra integrodifferential equations in [9], nonlinear thermoelasticity in [27], as well as for the compressible Navier–Stokes equations in [21] and [6] (see also the references cited in [17, 9, 27, 21, 6]).

Consider the following scalar model with damping:

$$u_t + f(u)_x + u = 0, \tag{2.1}$$

with Cauchy data

$$u|_{t=0} = u_0(x). \tag{2.2}$$

The local existence and uniqueness of the smooth solution $u(x,t)$ for the Cauchy problem (2.1)–(2.2) can be found in [19]. It can be obtained by using the standard Contraction Mapping Theorem of Banach (see Section 3.1 for such an application for the inviscid hydrodynamic energy model). In this section, we present the energy method, through the scalar model (2.1), in order to motivate the more general approach involved in establishing the global existence, uniqueness, and asymptotic behavior of classical smooth solutions to systems of conservation laws to follow in the next section. It is particularly intended for readers not familiar with the techniques of this subject.

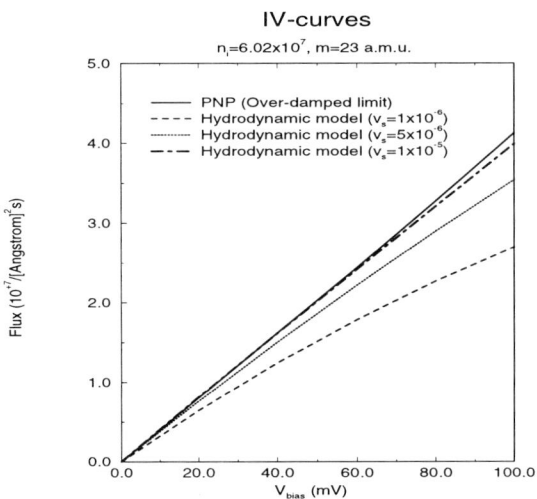

FIG. 3. The IV curves are calculated for different values of v_s and compared to the drift-diffusion (PNP) case. From [5]

Theorem 2.1 *Assume that initial data $u_0(x) \in H^2(\mathbb{R})$. Then there exists a unique global smooth solution of the problem (2.1)–(2.2) satisfying*

$$u(x,t) \to 0, \quad \text{uniformly when} \quad t \to \infty.$$

Proof It suffices to obtain the energy estimates for the local solution $u(x,t)$. Now we derive such estimates in considerable detail, as an illustration for the following Sections 3 and 4, for the smooth solution $u(x,t) \in H^2((-\infty,\infty) \times [0,t_0))$. In the course of these estimates, we make certain assumptions on f, not always made explicit. We first assume a priori that

$$\|u\|_{C^1(\mathbb{R}\times[0,t_0))} \leq \varepsilon, \tag{2.3}$$

for sufficiently small ε. Then we prove that there exists $\varepsilon_0 = \varepsilon_0(\varepsilon, f)$ such that, when

$$\|u_0\|_{H^2} \leq \varepsilon_0,$$

the a priori assumption (2.3) is indeed achieved.

Multiplying (2.1) by u, one has

$$\left(\frac{u^2}{2}\right)_t + u^2 + \left(\int_0^u vf'(v)dv\right)_x = 0.$$

Integrating over $\mathbb{R} \times [0, t)$, $t \leq t_0$ leads to

$$\int_{-\infty}^{\infty} u^2(x,t)dx + 2\int_0^t \int_{-\infty}^{\infty} u^2(x,s)dxds = \int_{-\infty}^{\infty} u_0^2(x)dx. \qquad (2.4)$$

Differentiating (2.1) with respect to x and then multiplying by u_x, one has

$$\left(\frac{u_x^2}{2}\right)_t + u_x^2 + \frac{1}{2}f''(u)u_x^3 = -\frac{1}{2}(f'(u)u_x^2)_x.$$

Integrating over $\mathbb{R} \times [0, t)$, $t \leq t_0$ leads to

$$\int_{-\infty}^{\infty} u_x^2(x,t)dx + 2\int_0^t \int_{-\infty}^{\infty} u_x^2(x,s)dxds$$
$$\leq \int_{-\infty}^{\infty} u_{0x}^2(x)dx + C \int_0^t \int_{-\infty}^{\infty} |u_x|^3(x,s)dxds. \qquad (2.5)$$

Differentiating (2.1) with respect to t and then multiplying by u_t, one has

$$\left(\frac{u_t^2}{2}\right)_t + u_t^2 + \frac{1}{2}f''(u)u_x u_t^2 = -\frac{1}{2}(f'(u)u_t^2)_x.$$

Integrating over $\mathbb{R} \times [0, t)$, $t \leq t_0$ leads to

$$\int_{-\infty}^{\infty} u_t^2(x,t)dx + 2\int_0^t \int_{-\infty}^{\infty} u_t^2(x,s)dxds$$
$$\leq \int_{-\infty}^{\infty} u_{0t}^2(x)dx + C\int_0^t \int_{-\infty}^{\infty} (|u_x|u_t^2)(x,s)dxds$$
$$\leq C\int_{-\infty}^{\infty} (u_0^2 + u_{0x}^2)(x)dx + C\int_0^t \int_{-\infty}^{\infty} (|u_x|u_t^2)(x,s)dxds. \qquad (2.6)$$

Here we have have used the notation $u_{0t}(x) = u_t(x,0)$, and estimated this quantity by (2.1). Differentiating (2.1) with respect to xx and then multiplying by u_{xx}, one has

$$\left(\frac{u_{xx}^2}{2}\right)_t + u_{xx}^2 + 2f''(u)u_x u_{xx}^2 + f'''(u)u_x^3 u_{xx} = -\frac{1}{2}(f'(u)u_{xx}^2)_x.$$

Integrating over $\mathbb{R} \times [0, t)$, $t \leq t_0$ leads to

$$\int_{-\infty}^{\infty} u_{xx}^2(x,t)dx + 2\int_0^t \int_{-\infty}^{\infty} u_{xx}^2(x,s)dxds$$

$$\leq \int_{-\infty}^{\infty} u_{0xx}^2(x)dx + C\int_0^t \int_{-\infty}^{\infty} (|u_x u_{xx}^2| + |u_x^3 u_{xx}|)(x,s)dxds$$

$$\leq C\int_{-\infty}^{\infty} u_{0xx}^2(x)dx + C\int_0^t \int_{-\infty}^{\infty} (|u_x u_{xx}^2| + u_x^6)(x,s)dxds, \tag{2.7}$$

where we used the inequality,

$$ab \leq \delta a^2 + b^2/(4\delta). \tag{2.8}$$

The left hand side of (2.7) is understood to be the expression two lines above; for compactness, we absorbed the estimation term δu_{xx}^2, and did not rewrite this left hand side. We employ this technique throughout the proof.

Differentiating (2.1) with respect to xt and then multiplying by u_{xt}, one has

$$\left(\frac{u_{xt}^2}{2}\right)_t + u_{xt}^2 + \frac{3}{2}f''(u)u_x u_{xt}^2 + f''(u)u_t u_{xx} u_{xt} + f'''(u)u_x^2 u_t u_{xt} = -\frac{1}{2}(f'(u)u_{xt}^2)_x.$$

Integrating over $\mathbb{R} \times [0,t]$, $t \leq t_0$ leads to

$$\int_{-\infty}^{\infty} u_{xt}^2(x,t)dx + 2\int_0^t \int_{-\infty}^{\infty} u_{xt}^2(x,s)dxds$$

$$\leq \int_{-\infty}^{\infty} u_{0xt}^2(x)dx + C\int_0^t \int_{-\infty}^{\infty} (|u_x u_{xt}^2| + |u_t u_{xx} u_{xt}| + |u_x^2 u_t u_{xt}|)(x,s)dxds$$

$$\leq C\int_{-\infty}^{\infty} (u_{0x}^2 + u_{0x}^4 + u_{0xx}^2)(x)dx$$

$$+ C\int_0^t \int_{-\infty}^{\infty} (|u_x u_{xt}^2| + u_t^2 u_{xx}^2 + u_x^4 u_t^2)(x,s)dxds, \tag{2.9}$$

where we used the inequality (2.8) and (2.1), and the notation, $u_{0xt} = u_{xt}(x,0)$.

Similarly, differentiating (2.1) with respect to tt and then multiplying by u_{xt}, one has

$$\left(\frac{u_{tt}^2}{2}\right)_t + u_{tt}^2 + \frac{1}{2}f''(u)u_x u_{tt}^2 + 2f''(u)u_t u_{xt} u_{tt} + f'''(u)u_x u_t^2 u_{tt} = -\frac{1}{2}(f'(u)u_{tt}^2)_x.$$

Integrating over $\mathbb{R} \times [0,t]$, $t \leq t_0$ leads to

$$\int_{-\infty}^{\infty} u_{tt}^2(x,t)dx + 2\int_0^t \int_{-\infty}^{\infty} u_{tt}^2(x,s)dxds$$

$$\leq \int_{-\infty}^{\infty} u_{0tt}^2(x)dx + C\int_0^t \int_{-\infty}^{\infty} (|u_x u_{tt}^2| + |u_t u_{xt} u_{tt}| + |u_x u_t^2 u_{tt}|)(x,s)dxds$$

$$\leq C \int_{-\infty}^{\infty} (u_{0xx}^2 + u_{0x}^4 + u_0^4 + u_{0x}^2 + u_0^2)(x)dx$$
$$+ C \int_0^t \int_{-\infty}^{\infty} (|u_x u_{tt}^2| + u_t^2 u_{xt}^2 + u_x^2 u_t^4)(x,s)dxds, \qquad (2.10)$$

where we used the inequality (2.8) and (2.1), and the notation, $u_{0tt} = u_{tt}(x,0)$. Adding (2.4)–(2.7) and (2.8)–(2.10), one has

$$\int_{-\infty}^{\infty} (u^2 + u_x^2 + u_t^2 + u_{xx}^2 + u_{xt}^2 + u_{tt}^2)(x,t)dx$$
$$+ 2 \int_0^t \int_{-\infty}^{\infty} (u^2 + u_x^2 + u_t^2 + u_{xx}^2 + u_{xt}^2 + u_{tt}^2)(x,s)dxds$$
$$\leq C \int_{-\infty}^{\infty} (u_0^2 + u_0^4 + u_{0x}^2 + u_{0x}^4 + u_{0xx}^2)(x)dx$$
$$+ C \int_0^t \int_{-\infty}^{\infty} [u_x^2 u_t^2(u_t^2 + u_x^2) + u_x^6 + |u_x|(u_x^2 + u_t^2 + u_{xx}^2 + u_{xt}^2 + u_{tt}^2)$$
$$+ u_t^2(u_{xx}^2 + u_{xt}^2)](x,s)dxds$$
$$\leq C\|u_0\|_{H^2(\mathbb{R})}^2 + C\varepsilon \int_0^t \int_{-\infty}^{\infty} (u_x^2 + u_t^2 + u_{xx}^2 + u_{xt}^2 + u_{tt}^2)dxds. \qquad (2.11)$$

For sufficiently small $\varepsilon > 0$, one has from (2.11) that

$$[N(t)]^2 = \int_{-\infty}^{\infty} (u^2 + u_x^2 + u_t^2 + u_{xx}^2 + u_{xt}^2 + u_{tt}^2)(x,t)dx$$
$$+ 2 \int_0^t \int_{-\infty}^{\infty} (u^2 + u_x^2 + u_t^2 + u_{xx}^2 + u_{xt}^2 + u_{tt}^2)(x,s)dxds$$
$$\leq C\|u_0\|_{H^2(\mathbb{R})}^2. \qquad (2.12)$$

Notice that C does not depend upon ϵ if the latter is sufficiently small. From the local existence and uniqueness theorem, there exists an interval $[0, t_0)$ such that (2.1) has a unique smooth solution. It follows from the Sobolev inequality that

$$\|u\|_{C^1(\mathbb{R} \times [0,t_0))} \leq \Gamma \sup_{0 \leq t \leq t_0} N(t),$$

where Γ is a positive constant. Choose ϵ_0 such that

$$e_0 \leq \frac{\varepsilon}{\Gamma\sqrt{C}}.$$

Then the estimate (2.12) implies

$$N(t) \leq \varepsilon/\Gamma,$$

as we assume a priori, which implies (2.3). By a technique developed fully in the next sections, we can apply the above arguments again, and continue the local solution to the global solution in $[0, \infty)$ by use of the estimate (2.12). Alternatively, if one wishes to use only the information already derived, one can assume that continuation has been carried out maximally, and then use the C^1 estimates to continue locally from classical conservation law principles. This contradiction ensures the global continuation.

Therefore, we have proved the existence of the global smooth solution u satisfying

$$u \in L^2([0, \infty); H^2(\mathbb{R})); \qquad \|u\|_{C(\mathbb{R})}(t) \to 0, \quad \text{when } t \to \infty.$$

This completes the proof. □

3 Global Smooth Solutions of the Inviscid Hydrodynamic Energy Model

In this section, we consider the inviscid hydrodynamic energy model, rewritten from (1.8) as:

$$\rho_t + (\rho v)_x = 0, \tag{3.1}$$

$$(\overline{m}\rho v)_t + (\overline{m}\rho v^2 + \rho T)_x = \overline{e}\rho\phi_x - \frac{\overline{m}\rho v}{\tau_p}, \tag{3.2}$$

$$E_t + (vE + v\rho T)_x - (\kappa T_x)_x = \overline{e}\rho v\phi_x - \frac{E - \frac{3}{2}\rho \overline{T}}{\tau_w}, \tag{3.3}$$

$$\phi_{xx} = \overline{e}(\rho - d(x)), \tag{3.4}$$

where $\rho(x, t) > 0$, $v(x, t)$, $T(x, t)$, $\phi(x, t)$, $\kappa > 0$, $d(x, t) \geq d_0 > 0$, $\overline{e} > 0$, $\overline{T}$, and $\overline{m}$ respectively denote *electron density*, *velocity*, *temperature* in energy units (the Boltzmann constant k_b has been set to 1), *electrostatic potential* in charge units per length (the dielectric has been set to 1), *thermal conductivity*, *doping profile* (background ion density), *electron charge*, *ambient device temperature*, and *effective electron mass*, and

$$E = \frac{3}{2}\rho T + \frac{1}{2}\overline{m}\rho v^2,$$

is the *energy density* for parabolic energy bands. The collision terms have been treated classically through momentum and energy relaxation times $\tau_p > 0$ and $\tau_w > 0$ (see the discussion in Section 1.2).

The existence of a global classical solution of the system (3.1)–(3.4) with additional viscosity was proved in [30]. However, for the Cauchy problem of (3.1)–(3.4), with large smooth initial data, it is proved in [29] that the solution generally develops a singularity, shock waves, and hence no global classical solution exists. The next question is whether there exists a global classical solution of

(3.1)–(3.4), provided that the initial data are small, although the system (3.1)–(3.4) does not have the viscosity term as in [30]. In this section, we address this problem with the aid of the energy method as shown in Section 2, and prove that there exists a global smooth solution of the inviscid system (3.1)–(3.4), provided that the initial data are small.

Consider the following initial-boundary value problem with impermeable insulated boundaries for (3.1)–(3.4):

$$(\rho, v, T)(x, 0) = (\rho_0(x), \tau_p v_0(x), \overline{T} + \tau_w(T_0(x) - \overline{T})), \quad 0 \le x \le 1,$$
$$(v, T_x, \phi)(i, t) = (0, 0, 0), \quad i = 0, 1, \quad t \ge 0, \tag{3.5}$$

Such an initial condition is natural for the drift-diffusion limit (see Section 4). For the existence problem, this initial condition can be relaxed by modifying our energy estimates. Note that the boundary conditions for ϕ are those of electric neutrality.

In this section we use the energy method to prove the global existence, uniqueness, and asymptotic decay of classical smooth solutions for the problem (3.1)–(3.5) without the viscosity terms. This indicates that the relaxation term prevents the development of shock waves for smooth initial data with small oscillation.

We remark that, in the hydrodynamic equations (3.1)–(3.4), we can assume $m = \overline{e} = 1$ without loss of generality. Otherwise we can rescale

$$(T, \overline{T}, \phi) \to (T/m, \overline{T}/m, \overline{e}\phi/m)$$

in the first three equations and the resultant constant in the fourth equation will not add any analytical difficulty, and hence can be assumed to equal one in our analysis.

Using the Lagrangian coordinate (y, t) as used in [6]: $y = \int_0^x \rho(\xi, t)d\xi$, we obtain the full hydrodynamic model in Lagrangian coordinates:

$$(1/\rho)_t - v_y = 0, \tag{3.6}$$
$$v_t + (\rho T)_y = \rho\phi_y - v/\tau_p, \tag{3.7}$$
$$T_t - \frac{2}{3}(\kappa\rho T_y)_y + \frac{2}{3}\rho T v_y = \frac{2\tau_w}{3\tau_p\tau_w}v^2 - \frac{T}{\tau_w}\frac{\overline{T}}{\tau_w}, \tag{3.8}$$
$$(\rho\phi_y)_y = 1 - \tilde{d}/\rho, \tag{3.9}$$

with the same initial-boundary conditions:

$$(\rho, v, T)(y, 0) = (\rho_0(x(y, 0)), \tau_p v_0(x(y, 0)), \overline{T} + \tau_w(T_0(x(y, 0)) - \overline{T})),$$
$$(v, T_y, \phi)(i, t) = (0, 0, 0), \quad i = 0, 1, \quad t \ge 0, \tag{3.10}$$

where $\tilde{d}(y, t) = d(x(y, t))$.

Let

$$1/\rho - 1 = w_y, \quad v = w_t, \quad \theta = T - \overline{T}. \tag{3.11}$$

Then, the system (3.6)–(3.9) is transformed into the following form:

$$w_{tt} - \frac{\theta+\overline{T}}{(1+w_y)^2}w_{yy} + \frac{1}{1+w_y}\theta_y + \frac{w_t}{\tau_p} = \int_0^1 \int_0^y ((1-\tilde{d}) - \tilde{d}w_y)d\xi d\eta, \quad (3.12)$$

$$\theta_t - \frac{2\kappa}{3(1+w_y)}\theta_{yy} + \frac{2\kappa\theta_y}{3(1+w_y)^2}w_{yy} + \frac{2(\theta+\overline{T})}{3(1+w_y)}w_{yt} + \frac{\theta}{\tau_w} - \frac{(2\tau_w - \tau_p)w_t^2}{3\tau_p \tau_w}$$
$$= 0, \quad (3.13)$$

with the following initial-boundary conditions:

$$(w, w_t, \theta)(y, 0)$$
$$= \left(\int_0^y \frac{1-\rho_0(x(\xi,0))}{\rho_0(x(\xi,0))}d\xi, \tau_p v_0(x(y,0)), \overline{T} + \tau_w(T_0(x(y,0)) - \overline{T})\right), (3.14)$$
$$(w, \theta_y)(i, t) = (0, 0), \quad i = 0, 1, \quad t \geq 0. \quad (3.15)$$

In the following context, we use the following notations for simplicity:

$$\partial = \partial^1 = (\partial_t, \partial_y), \quad \partial^2 = (\partial_{tt}^2, \partial_{yt}^2, \partial_{yy}^2), \quad \partial^3 = (\partial_{ttt}^3, \partial_{ytt}^3, \partial_{yyt}^3, \partial_{yyy}^3),$$
$$D = D^1 = (\partial^0, \partial^1), \quad D^2 = (\partial^0, \partial^1, \partial^2), \quad D^3 = (\partial^0, \partial^1, \partial^2, \partial^3),$$
$$|\partial^0 u|^2 = |u|^2, \quad |\partial^1 u|^2 = |\partial_t u|^2 + |\partial_y u|^2, \quad |\partial^2 u|^2 = |\partial_{tt} u|^2 + |\partial_{ty} u|^2 + |\partial_{yy} u|^2,$$
$$|\partial^3 u|^2 = |\partial_{ttt} u|^2 + |\partial_{tty} u|^2 + |\partial_{tyy} u|^2 + |\partial_{yyy} u|^2,$$

$$|D^m u|^2 = \sum_{j=0}^m |\partial^j u|^2, \quad m = 1, 2, 3,$$

$C > 0$ is a universal constant, and $C(N) > 0$ is a universal constant depending only on N.

3.1 Local Solutions

To establish the local existence for the problem (3.1)–(3.5), we apply Banach's contraction mapping theorem. For this purpose, we first consider the following linear problem:

$$w_{tt} - a_1(y,t)w_{yy} + a_3(y,t)\theta_y + w_t/\tau_p = f, \quad (3.16)$$
$$\theta_t - a_2(y,t)\theta_{yy} + a_4(y,t)w_{yy} + a_5(y,t)w_{yt} + a_6(y,t)w_t + \theta/\tau_w = 0, \quad (3.17)$$
$$(w, w_t, \theta)(y, 0) = (w_0, w_1, \theta_0)(y), \quad (w, \theta_y)(i, t) = (0, 0), \quad i = 0, 1. (3.18)$$

Lemma 3.1 *For $t_0 > 0$ and $1 \leq j \leq 6$, let*

$$a_1, a_2 > 0, \quad a_j(y,t) \in C^1([0,1] \times [0,t_0]), \quad \partial^2 a_j(y,t) \in L^\infty([0,t_0]; L^2(0,1)).$$

Assume that the initial data satisfy

$$(w_{0y}, w_1)(y) \in H^2(0,1), \quad \theta_0(y) \in H^3(0,1); \quad (w_0, w_1, \theta_{0y})(i) = (0,0,0), \quad i = 0, 1.$$

Then, (3.16)–(3.18) has a unique smooth solution (w, θ) satisfying:

$$w(y,t) \in C^2([0,1] \times [0,t_0]), \quad (\theta, \partial\theta_y)(y,t) \in C^1([0,1] \times [0,t_0]),$$
$$\theta_{tt}(y,t) \in L^\infty([0,t_0]; C[0,1]), (D^3w, D^2\theta, \partial^2\theta_y)(y,t) \in L^\infty([0,t_0]; L^2(0,1)).$$

Lemma 3.1 can be proved by applying the method of Faedo/Galerkin and Lions. For details, see Section 8.2 of Chapter 3 in [18]. One could also use Rothe's method of horizontal lines.

To establish the local existence of the initial-boundary problem (3.12)–(3.15), we modify the system (3.12)–(3.13) to avoid possible singular coefficients. Let $a(w_y)$ be a smooth function so that $a(w_y) = \frac{1}{1+w_y}$ for $|w_y| \leq \frac{1}{2}$ and $\frac{2}{3} \leq a(w_y) \leq 2$ for $|w_y| < \infty$. We also assume that $\theta + \overline{T} \geq \frac{\overline{T}}{2}$. This will cause no difficulty since we will have an a posteriori estimate with small $|w_y(y,t)|$ for small $t_0 > 0$.

Theorem 3.2 *Assume that initial data satisfy*

$$(\rho_0(y) - 1, v_0(y)) \in H^2(0,1), \ \theta_0(y) \in H^3(0,1); \quad (v_0, \theta_{0y})(i) = (0,0), \ i = 0, 1.$$

Then, the problem (3.12)–(3.15) has a unique smooth solution (w, θ) satisfying

$$w(y,t) \in C^2([0,1] \times [0,t^*)), \quad (\theta, \partial\theta_y)(y,t) \in C^1([0,1] \times [0,t^*)),$$
$$(D^3w, D^2\theta)(y,t) \in L^\infty([0,t^*); L^2(0,1)), \ (\partial^2\theta_y)(y,t) \in L^2([0,t^*); L^2(0,1)),$$

defined on a maximal interval of existence $[0, t^*)$. If $t^* < \infty$, then

$$\int_0^1 |(D^3w, D^2\theta)|^2 dy + \int_0^t \int_0^1 |\partial^2\theta_y|^2 dy ds \to \infty, \quad t \to t^*. \quad (3.19)$$

Proof *Step 1* For positive constants t_0 and N with

$$N^2 \geq 2\{\|(w_0, \theta)\|_{H^3(0,1)}^2 + \|w_{0t}\|_{H^2(0,1)}^2\},$$

we define

$$X(N, t_0) = \left\{ (U, \Theta)(y,t) : \begin{array}{l} \|(U,\Theta)\|_{X(N,t)} \leq N, \\ (U, U_t, \Theta)(y,0) = (w_0, \tau_p v_0, \overline{T} + \tau_w(T_0 - \overline{T}))(y), \\ (U, \Theta_y)(i,t) = (0,0), 0 \leq t \leq t_0, i = 0,1, \end{array} \right\}$$

where

$$\|(U,\Theta)\|_{X(N,t)}^2 = \sup_{0 \leq t \leq t_0} \int_0^1 |(D^3U, D^2\Theta)|^2 dy + \int_0^t \int_0^1 |\partial^2\Theta_y|^2 dy ds.$$

Let $(U, \Theta) \in X(N, t_0)$. Consider the following linear system (3.16)–(3.17) with initial-boundary data

$$(w, w_t, \theta)(y,0) = (w_0, \tau_p v_0, \overline{T} + \tau_w(T_0 - \overline{T}))(y), \ 0 \leq y \leq 1, \quad (3.20)$$
$$(w, \theta_y)(i,t) = (0,0), \ 0 \leq t \leq t_0, \ i = 0,1, \quad (3.21)$$

where

$$a_1(y,t) = (\Theta + \overline{T})a(U_y), \quad a_2(y,t) = 2\kappa a(U_y)/3, \quad a_3(y,t) = a(U_y),$$

$$a_4(y,t) = \frac{2\kappa\Theta_y}{3}a^2(U_y), \; a_5(y,t) = \frac{2(\Theta + \overline{T})}{3}a(U_y), \; a_6(y,t) = \frac{(\tau_p - 2\tau_w)U_t}{3\tau_p\tau_w}.$$

(3.22)

All coefficients of (3.22) in the equations (3.16)–(3.17) satisfy the conditions of Lemma 3.1. Thus, for every $(U, \Theta) \in X(N, t_0)$, the problem (3.16)–(3.17) and (3.20)–(3.21) has a unique solution $(w, \theta) \in X(M, t_0)$ for certain $M \geq N > 0$. Now we define a map F by solving (3.16)–(3.17) and (3.20)–(3.21), i.e., $(w, \theta) = F(U, \Theta)$.

Step 2 To show that there exists $\tilde{t} \leq t_0$ such that F maps $X(N, \tilde{t})$ into itself, we need a priori estimates. Since $(w, \theta) \in X(M, t_0), M \geq N > 0$, it suffices to prove that there exists $\tilde{t} \leq t_0$ such that $(w, \theta) \in X(N, \tilde{t})$.

Multiply (3.16) and (3.17) by w_t and θ, respectively. Differentiate (3.16), (3.17) with respect to t and multiply by w_{tt}, θ_t, respectively. Differentiate (3.16), (3.17) with respect to y and multiply by w_{yt}, θ_y, respectively. Then we differentiate (3.16), (3.17) with respect to tt and multiply by w_{ttt}, θ_{tt}, respectively. Also we take second derivatives of (3.16) and (3.17) with respect to yt and multiply by w_{ytt}, θ_{yt}, respectively; and take derivatives of (3.16) and (3.17) with respect to yy and y, and multiply by $a_1 w_{yyt} + f$, θ_{yyy}, respectively. Finally we take derivatives of (3.16) with respect to yy and multiply by $(a_1 w_{yyt} + f)_t$. We then use integration by parts and the elementary inequality: $\alpha\beta \leq \frac{1}{2}(\alpha^2 + \beta^2)$, to obtain the following energy inequality:

$$\frac{1}{2}\int_0^1 \left[w_t^2 + a_1|D^2 w_y|^2 + |D^2 w_t|^2 + |D^2\theta|^2\right] dy$$

$$+ \int_0^t \int_0^1 \left[\frac{1}{\tau_p}|D^2 w_t|^2 + \frac{1}{\tau_w}|D^2\theta|^2 + a_2|D^2\theta_y|^2\right] dyds$$

$$\leq I(0) + C\int_0^t \int_0^1 \Bigg[|D(Dw, \theta)|\bigg(\sum_{j=1}^{6}|\partial^2 a_j| + |\partial a_4|\bigg)|D^2(Dw, \theta, \theta_y)|$$

$$+ |D^2 f \cdot D^2 w_t| + \bigg(\sum_{j \neq 4}(|a_j| + |\partial a_j|) + |a_4|\bigg)|D^2(Dw, \theta, \theta_y)|^2\Bigg] dyds, \quad (3.23)$$

where $I(0)$ is the value of the first line above, in (3.23), evaluated at $t = 0$.

Since $(U, \Theta) \in X(N, t_0)$ and $(w, \theta) \in X(M, t_0)$, it follows from Sobolev's inequality that

$$\|(D^2 U, D\Theta)\|_{C([0,1]\times[0,t_0))} \leq CN, \quad \|(D^2 w, D\theta)\|_{C([0,1]\times[0,t_0))} \leq CM.$$

Then,

$$\int_0^t \int_0^1 |a_{1y}|(w_t^2 + w_y^2) dy ds$$
$$= \int_0^t \int_0^1 |\Theta_y a(U_y) + (\Theta + \overline{T})a'(U_y)U_{yy}|(w_t^2 + w_y^2) dy ds$$
$$\leq C(N) \int_0^t \int_0^1 (w_t^2 + w_y^2) dy ds \leq C(N) M^2 t.$$

Every term on the right hand side of (3.20) involving $a_4, a_j, \partial a_j, j \neq 4$, can be handled in a similar way.

To estimate the other terms we use a slightly different technique. For example, consider

$$\int_0^t \int_0^1 |a_{5tt}| |\theta_{tt}| |w_{yy}| dy ds,$$

where

$$a_{5tt} = \frac{2\kappa}{3}\Theta_{ytt} a^2 + 4\Theta_{yt} aa' U_{yt} + 2\Theta_y (a')^2 U_{yt}^2 + 2\Theta_y aa'' U_{yt}^2 + 2\Theta_y aa' U_{ytt}.$$

Since $(U, \Theta) \in X(N, t_0)$ and $(w, \theta) \in X(M, t_0)$, we have

$$\int_0^t \int_0^1 |a_{5tt} \theta_{tt} w_{yy}| dy ds$$
$$\leq C(N) \left(\int_0^t \int_0^1 |\theta_{tt} w_{yy}|^2 dy ds \right)^{1/2} \|D^3(U, \Theta)\|_{L^2([0,1]\times[0,t_0])}$$
$$\leq C(N) NM \left(\int_0^t \int_0^1 w_{tt}^2 dy ds \right)^{1/2} \leq \sqrt{t} C(N) MN.$$

Similarly, we can estimate the terms involving the second derivatives of a_j.

Then, it follows from (3.23) that

$$\int_0^1 (|D^3 w|^2 + |D^2 \theta|^2) dy + \int_0^t \int_0^1 (|D^2 w_t|^2 + |D^2 \theta|^2 + |D^2 \theta_y|^2) dy ds$$
$$\leq I(0) + M^2 C(N)(t + \sqrt{t}) \leq \frac{N^2}{2} + M^2 C(N)(t + \sqrt{t}). \tag{3.24}$$

Now choose $\tilde{t}$ sufficiently small so that $M^2 C(N)(t + \sqrt{t}) \leq \frac{N^2}{2}$. We then have $(w, \theta) \in X(N, \tilde{t})$.

Step 3 Now we show F is a contraction map in $X(N, \tilde{t})$ for sufficiently small $\tilde{t} \leq 1$.

Let $(U, \Theta), (\widehat{U}, \widehat{\Theta}) \in X(N, t_0)$. Then $(w, \theta) = F(U, \Theta)$, $(\widehat{w}, \widehat{\theta}) = F(\widehat{U}, \widehat{\Theta})$, and $(w - \widehat{w}, \theta - \widehat{\theta})$ satisfies

$$(w - \widehat{w})_{tt} - a_1(w - \widehat{w})_{yy} + a_3(\theta - \widehat{\theta})_y + \frac{1}{\tau_p}(w - \widehat{w})$$
$$= (f - \widehat{f}) + (a_1 - \widehat{a_1})\widehat{w}_{yy} - (a_3 - \widehat{a_3})\widehat{\theta}_y,$$
$$(\theta - \widehat{\theta})_t - a_2(\theta - \widehat{\theta})_{yy} + a_4(w - \widehat{w})_{yy} + a_5(w - \widehat{w})_{yt} + a_6(w - \widehat{w})_t + \frac{1}{\tau_w}(\theta - \widehat{\theta})$$
$$= (a_2 - \widehat{a_2})\widehat{\theta}_{yy} - (a_4 - \widehat{a_4})\widehat{w}_{yy} - (a_5 - \widehat{a_5})\widehat{w}_{yt} - (a_6 - \widehat{a_6})\widehat{w}_t.$$

In a manner similar to deriving the energy estimates required for (3.24), and with the aid of Gronwall's inequality, we can obtain

$$\|(w - \widehat{w}, \theta - \widehat{\theta})\|_{X(N,\tilde{t})} \leq \tilde{t}C(N)\|(U - \widehat{U}, \Theta - \widehat{\Theta})\|_{X(N,\tilde{t})}.$$

Choose $\tilde{t} \leq 1$ sufficiently small such that $0 < \tilde{t}C(N) < 1$. Thus

$$F : X(N, \tilde{t}) \to X(N, \tilde{t})$$

is a contraction mapping on $X(N, \tilde{t})$ for sufficiently small $\tilde{t} > 0$.

The final step is to apply Banach's Contraction Mapping Theorem. We see that F has a unique fixed point in $X(N, \tilde{t})$. That is, (3.14)–(3.15) has a unique smooth solution (w, θ) on $0 \leq t \leq \tilde{t}$. Let $[0, t^*)$ denote the maximal interval of existence of this solution. If $t^* < \infty$ and (3.19) is not satisfied, then

$$\sup_{0 \leq t \leq t^*} \int_0^1 (|D^3 w|^2 + |D^2 \theta|^2) dy + \int_0^t \int_0^1 |\partial^2 \theta_y|^2 dy ds < \infty.$$

This inequality and use of the second equation in (3.16) imply that $(w, \theta)(y, t^*)$ satisfies the conditions for initial data in Theorem 3.2. Hence, the local existence can be continued to the interval $[0, t^* + \varepsilon)$ for some $\varepsilon > 0$. This contradicts the fact that $[0, t^*)$ is the maximal interval of existence. This completes the proof of Theorem 3.2. □

3.2 A Priori Estimates via the Energy Method

In this section we use the energy method to make the a priori estimates for establishing the global existence of smooth solutions for the full hydrodynamic model (3.1)–(3.4). To achieve this, we derive the a priori estimates for the solution of (3.12)–(3.15), which is equivalent to (3.1)–(3.5).

Theorem 3.3 *Assume that*

$$0 < \delta \leq \rho_0(x) \leq M, \qquad (v_0, \theta_{0x})(i) = (0, 0), \ i = 0, 1,$$
$$|2\tau_w - \tau_p| \leq M\sqrt{\tau_w \tau_p}.$$

Then, given sufficiently small $\varepsilon > 0$, there exists $\varepsilon_0(\varepsilon) > 0$ such that, when

$$\|(\rho_0 - 1, v_0, d - 1)\|_{H^2(0,1)} + \|T_0 - 1\|_{H^3(0,1)} + \tau_p \leq \epsilon_0,$$

we have, for any time interval $(0, t_0)$ on which the solution exists,

$$\|(D^2w, D^2w_y, D\theta, D\theta_y, \tau_p w_{ttt}, \tau_p \theta_{tt})\|_{L^2(0,1)}(t) + \|(\theta_y, \partial\theta_{yy})\|_{L^2((0,1)\times(0,t_0))}$$
$$+ \frac{1}{\tau_p}\|(Dw_t, \partial w_{yy}, D\theta, D\theta_y)\|_{L^2((0,1)\times(0,t_0))}$$
$$+ \tau_p\|(\partial_y^2 w_y, w_{ttt}, D_y\theta_{tt})\|_{L^2((0,1)\times(0,t_0))} \leq \varepsilon. \quad (3.25)$$

Proof We start with the following equations from (3.12)–(3.13):

$$w_{tt} - a(w_y)^2(\theta + \overline{T})w_{yy} + a(w_y)\theta_y + \frac{w_t}{\tau_p} + w - \int_0^1 w\,dy = f, \quad (3.26)$$

$$\theta_t - \alpha a(w_y)\theta_{yy} + \alpha a(w_y)^2\theta_y w_{yy} + \beta a(w_y)(\theta + \overline{T})w_{yt} + \frac{\theta}{\tau_w} + b\frac{w_t^2}{\sqrt{\tau_p \tau_w}} = 0, \quad (3.27)$$

with the initial-boundary conditions (3.14)–(3.15), where

$$f(y,t) = \int_0^1 \int_\eta^y (1-\tilde{d})(1+w_y)\,d\xi\,d\eta, \quad (3.28)$$

and

$$a(w_y) = \frac{1}{1+w_y}, \quad b = -\frac{2\tau_w - \tau_p}{3\sqrt{\tau_p \tau_w}}, \quad \alpha = 2\kappa/3, \quad \beta = 2/3.$$

Then, from the assumptions of Theorem 3.3,

$$|b| \leq C.$$

We first assume that, for sufficiently small ε,

$$\|(w, w_y, \theta)\|_{C^1([0,1]\times[0,t_0))} \leq \varepsilon \leq \frac{1}{2}\min(1, \overline{T}), \quad (3.29)$$

which implies

$$\overline{T}/2 \leq T \leq 3\overline{T}/2,$$

and

$$1/2 \leq w_y + 1 \leq 3/2 \iff 2/3 \leq \rho(x,t) \leq 2.$$

Then we prove that there exists an $\epsilon_0 > 0$, independent of t_0, such that, when

$$\|(\rho_0 - 1, v_0, d - 1)\|_{H^2(0,1)} + \|T_0 - 1\|_{H^3(0,1)} + \tau_p \leq \epsilon_0,$$

the estimates (3.29) can be achieved. With this a priori argument, $a(w_y)$ is a C^∞ function of one variable w_y such that $|w_y| \leq \varepsilon < 1/2$ and $2/3 \leq a(w_y) \leq 2$.

We have from (3.26) and (3.14)–(3.15) that

$$((\theta + \overline{T})a^2 w_{yy} + f + \int_0^1 w\,dy)(i,t) = 0, \quad i = 0, 1. \quad (3.30)$$

We now derive the energy estimates. We first perform the following calculations: $(3.26) \times (\tau_p K^2 w)$, $(3.26) \times (K^2 w_t)$, $(3.26) \times (-\tau_p w_{yy})$, $(3.26)_t \times (K^2 w_{tt})$, $(3.26)_y \times (K^2 w_{yt})$, $(3.26)_{tt} \times (\tau_p^2 w_{ttt})$, $(3.26)_{yt} \times w_{ytt}$, $(3.26)_{yy} \times (\tau((\theta+\overline{T})a^2 w_{yy} + f + \int_0^1 w dy))$, $(3.26)_{yy} \times (K((\theta+\overline{T})a^2 w_{yy} + f + \int_0^1 w dy)_t)$, $(3.27) \times (K^2 \theta)$, $(3.27) \times \theta_t$, $(3.27)_y \times \theta_y$, $(3.27)_{tt} \times (\tau_p^2 \theta_{tt})$, $(3.27)_{yt} \times \theta_{yt}$, and $(3.27)_y \times \theta_{yyy}$; then integrate over $[0,1] \times [0,t]$, where $K = K(\gamma, k) \geq 1$ is determined later. Then we integrate by parts with the aid of the boundary conditions (3.15) and (3.30), followed by use of some elementary inequalities such as

$$ab \leq \delta a^2 + \frac{b^2}{4\delta}, \quad |w| = \left|\int_0^1 w_y dy\right| \leq \left(\int_0^1 w_y^2 dy\right)^{1/2}. \quad (3.31)$$

Finally, we add and re-group all the resulting inequalities from previous calculations and estimates together, and use $\tau_p \leq 1/2$ to obtain:

$$\int_0^1 [K^2|(D^2 w, \theta)|^2 + K\left(((\theta+\overline{T})a^2 w_{yy} + f)_y^2 + \left(w_{yy} + \frac{f + \int_0^1 w dy}{(\theta+\overline{T})a^2}\right)_t^2\right)$$
$$+ |D(w_{yt}, \theta, \theta_y)|^2 + \tau_p^2(w_{ttt}^2 + \theta_{tt}^2)](y, t) dy$$
$$+ \int_0^t \int_0^1 [K^2(\tau_p w_y^2 + \theta_y^2 + \tau_p^{-1}|(Dw_t, w_{ytt}, \theta)|^2) + \tau_p^{-1}(K w_{yyt}^2 + |\partial(\theta, \theta_y)|^2)$$
$$+ \tau_p(w_{yy}^2 + \theta_{tt}^2 + |\partial w_{tt}|^2 + ((\theta+\overline{T})a^2 w_{yy} + f)_y^2) + |\partial \theta_{yy}|^2 + \tau_p \theta_{ytt}^2] dy ds$$
$$\leq C \int_0^1 \{K^2|(D^2 w, \theta)|^2 + K[((\theta+\overline{T})a^2 w_{yy} + f)_y^2 + \left(w_{yy} + \frac{f}{(\theta+\overline{T})a^2}\right)_t^2$$
$$+ (|\theta_t| + |w_{yt}|)(|w_{yy}| + |f| + \int_0^1 |w| dy)(|w_{yyt}| + |f_t|)$$
$$+ \int_0^1 |w_t| dy + (|f| + \int_0^1 |w| dy)(|\theta_t| + |w_{yt}|))]$$
$$+ |D(w_{yt}, \theta, \theta_y)|^2 + \tau_w^2 \theta_{tt}^2 + |D_t f|^2 + \tau_p^2 w_{ttt}^2 + w_{yy}^2(\theta_t^2 + w_{yt}^2)\}(y, 0) dy$$
$$+ C \int_0^1 [\tau_p^2 K^2 w_t^2 + K w_{yt}^2((\theta+\overline{T})a^2 w_{yy} + f + \int_0^1 w dy)_y^2 + w^2$$
$$+ \tau_p^2(w_{yt}^2 + w_{yyt}^2) + f_t^2 + w_{yt}^2 w_{yy}^2 + \tau_p |w_{yyt} f|$$
$$+ \tau_p^2(\theta_t^2 w_{yy}^2 + \theta_{yt}^2 w_{yy}^2 + \theta_y^2 w_{yyt}^2 + w_{yt}^2 \theta_y^2 w_{yy}^2)] dy$$
$$+ C \int_0^t \int_0^1 \{K^2[\tau_p(|(D^2 w_t, D^2 \theta_y, \theta_{yt})|^2 + \tau_p|\partial_y f|^2 + (w^2 + w_y^2 + w_{yt}^2) w_{yy}^2)$$
$$+ w_y^2 |w_{yt}| + w_t^4 + \tau_p w_{yy}^4 + \theta_y^2|(w, Dw_y, \partial_y^3 w)|^2$$
$$+ (|\theta_t| + |w_{yt}|) w_{yt}^2 + |\theta|(|w_y w_{yt}| + |w_{yy} w_{yyt}|)]$$
$$+ K[\tau_p^{-1}(w_t^2 + f^2 + \tau_p f_{tt}^2 + w_{yy}^2(w_{yt}^2 + \theta_t^2) + (f + \int_0^1 w dy)^2(w_{yt}^2 + \theta_t^2))$$

$$+ \tau_p w_{yt}^2 (f_t^2 + \int_0^1 w_t^2 dy) + |w_{yy}|(|f_t| + \int_0^1 |w_t| dy)$$

$$+ (|w_{yt}| + |\theta_t|)(w_{yyt}^2 + w_{yt}^2 w_{yy}^2 + f_t^2 + \int_0^1 |w_t|^2 dy + w_{yt}^2 + \theta_t^2)$$

$$+ |(\theta + \overline{T})a^2 w_{yy} + f + \int_0^1 w dy|(\tau_p(w_{ytt}^2 + w_{yt}^4 + \theta_{tt}^2 + \theta_t^2 w_{yt}^2)$$

$$+ \tau_p^{-1}(w_{yyt}^2 + w_{yt}^2 w_{yy}^2 + f_t^2 + w_t^2 + (f + \int_0^1 w dy)^2 (w_{yt}^2 + \theta_t^2))]$$

$$+ \tau_p w^2 + \tau_p w_{yt}^2 + |Dw_{yt}|^2 + \tau_p \theta_{yyy}^2 + \tau_p^2 w_{ttt}^2 + \tau_p^3 f_{tt}^2$$

$$+ \tau_p^3 (w_{yy}^2 |Dw_{tt}|^2 + w_{yt}^2 (w_y^4 + w_{yy}^2 w_{yt}^2 + w_{ytt}^2)) + \tau_p w_{yyt}^2 |Dw_y|^2$$

$$+ \tau_p^2 (w_{tt}^4 + w_t^2 w_{ttt}^2) + w_{yt}^2 (w_{yy}^2 + w_{tt}^2) + w_t^2 w_{ytt}^2$$

$$+ \theta_y^2 [\tau_p^3 (w_{ytt}^2 + w_{yt}^4 + \theta_y^2 w_{ttt}^2) + w_{yt}^2 + w_{yyt}^2 + w_{yyy}^2 + w_{yy}^4 + w_{yt}^2 w_{yy}^2]$$

$$+ \theta_t^2 [\tau_p^3 (w_{yyt}^2 + w_{ytt}^2 + w_{yy}^2 w_{yt}^2 + w_{yt}^4) + w_{yt}^2] + \tau_p^3 |\theta_t| w_{ytt}^2$$

$$+ w_{yy}^2 [\tau_p^3 (\theta_{tt}^2 + \theta_{ytt}^2) + \tau_p (\theta_{yt}^2 + \theta_{yy}^2 + \theta_{yyt}^2 + \theta_{yyy}^2))]$$

$$+ w_{yt}^2 [\tau^3 (\theta_{yt}^2 + \theta_{yyt}^2 + w_{yt}^2 \theta_{yy}) + \theta_{yy}^2] + \tau_w^3 \theta_{yy}^2 w_{ytt}^2$$

$$+ (|w_{yt}| + |\theta_t|) w_{yyt}^2 + \tau_w^2 |\theta_{ttt}| (|\theta_y w_{yyt}| + |\theta_{yt} w_{yy}| + |w_{yt} w_{yy} \theta_y|)$$

$$+ |w_{yy} w_{yyt}|(|\theta| + \tau_p |\theta_t|) + |w_{yyt}|(|w_{yy}| w_{yt}^2 + |f_{tt}| + |w_{yy} \theta_{tt}|)$$

$$+ \tau_p^2 |w_{yt}| w_{ytt}^2 + (|w_{yy}| + \tau_p |w_{yyt}|)(|f_t| + \int_0^1 |w_t| dy)$$

$$+ \tau_p^{-1} (w_{yt}^2 |Dw_y|^2 + w_t^2 |(Dw_t, w_{yyt})|^2) \} dy ds. \qquad (3.32)$$

Using the elementary inequalities (3.31) and the a priori assumption (3.29), taking $K \geq K_0(C, \overline{T})$ and $\tau_p \leq \tau_0(C, \overline{T}, K)$ for sufficiently large $K_0 \geq 1$ and small $\tau_0 \leq 1/2$, and making tedious calculations, we have from (3.32):

$$E_1(t) + \int_0^t G_1(s) ds \leq CK^2(\|(\rho_0 - 1, v_0, d_1)\|_{H^2(0,1)}^2 + \|T_0 - 1\|_{H^3(0,1)}^2)$$

$$+ \varepsilon C K^2 (E_1(t) + \int_0^t G_1(s) ds), \qquad (3.33)$$

where

$$E_1(t) = \int_0^1 [K^2 |(D^2 w, \theta)|^2 + K |Dw_{yy}|^2 + |D(w_{yt}, \theta, \theta_y)|^2 + \tau_p^2 (w_{ttt}^2 + \theta_{tt}^2)] dy,$$

$$G_1(t) = \int_0^1 [K^2 (\tau_p w_y^2 + \theta_y^2 + \frac{1}{\tau_p} |(Dw_t, w_{ytt}, \theta)|^2) + \tau_p (w_{yy}^2 + \theta_{tt}^2 + |(\partial_t w_{tt}, w_{yyy})|^2)$$

$$+ \tau_p^{-1} (K w_{yyt}^2 + |\partial(\theta, \theta_y)|^2) + |\partial \theta_{yy}|^2 + \tau_p \theta_{ytt}^2] dy ds. \qquad (3.34)$$

In the process of these calculations, we used the following identities:

$$\tilde{d}_t = -w_t d_x, \quad \tilde{d}_y = d_x/\rho, \quad \tilde{d}_{tt} = -w_t^2 d_{xx} - w_{tt} d_x,$$

and the following inequalities:

$$|f_y| \leq |1 - \tilde{d}| + |(\tilde{d} - 1)w_y|, \qquad |f_{yt}| \leq |w_{yt}(\tilde{d} - 1)| + |(w_y + 1)w_t d_x|,$$

$$|f_t| \leq \int_0^1 |f_{yt}| dy + \int_0^1 |w_{yt}| dy \int_0^1 |f_y| dy + \int_0^1 |w_y + 1| dy \int_0^1 |f_{yt}| dy,$$

$$|f_{tt}| \leq (2 + \int_0^1 |w_y| dy) \int_0^1 (|w_y + 1|(w_t^2|\tilde{d}_{xx}| + |w_{tt} \tilde{d}_x|) + 2|w_{yt} w_t \tilde{d}_x|$$

$$+ |\tilde{d} - 1||w_{ytt}|) d\eta + \int_0^1 |w_{ytt}| \int_0^1 |f_y| dy + 2 \int_0^1 |w_{yt}| \int_0^1 |f_{yt}| dy.$$

Then, if ϵ is sufficiently small, we have

$$E(t) + \int_0^t G(s) ds \leq C(\|(\rho_0 - 1, v_0, d - 1)\|_{H^2(0,1)}^2 + \|T_0 - 1\|_{H^3(0,1)}^2), \quad (3.35)$$

where

$$E(t) = \int_0^1 \left(|D^2(w, w_y)|^2 + |D(\theta, \theta_y)|^2 + \tau_p^2(w_{ttt}^2 + \theta_{tt}^2)\right) dy,$$

$$G(t) = \int_0^1 \left(\frac{1}{\tau_p}(|(Dw_t, \partial w_{yy})|^2 + |D(\theta, \theta_y)|^2)\right.$$

$$\left. + \tau_p |(\partial_y^2 w_y, w_{ttt})|^2 + |(\theta_y, \partial \theta_{yy})| + \tau_p |\partial_y \theta_{tt}|^2\right) dy. \quad (3.36)$$

By the Sobolev inequality we obtain

$$\|D(Dw, \theta)\|_{C([0,1]\times[0,t_0))} \leq \Gamma \sup_{0 \leq t \leq t_0} \sqrt{E(t)},$$

where Γ is a positive constant. Hence if

$$\|(\rho_0 - 1, v_0, d_0 - 1)\|_{H^2(0,1)} + \|\theta_0\|_{H^3(0,1)} \leq \frac{\varepsilon}{\Gamma\sqrt{C}}, \quad (3.37)$$

we have

$$\|D(Dw, \theta)\|_{C([0,1]\times[0,t_0))} \leq \Gamma\sqrt{C}(\|(\rho_0 - 1, v_0, d_0 - 1)\|_{H^2(0,1)} + \|\theta_0\|_{H^3(0,1)}) \leq \varepsilon$$

and the assumption for the function $a(w_y)$ is satisfied. This completes the proof of Theorem 3.3. □

3.3 Global Existence, Uniqueness, and Asymptotic Decay

Theorem 3.4 *Assume that*

$$0 < \delta \le \rho_0(x) \le M, \quad (v_0, \theta_{0x})(i) = (0,0), \ i = 0, 1, \quad |2\tau_w - \tau_p| \le M\sqrt{\tau_w \tau_p}.$$

Also, assume that the following norms,

$$\|(\rho_0 - 1, v_0, d - 1)\|_{H^2(0,1)}, \quad \|\theta_0\|_{H^3(0,1)},$$

and both relaxation times are sufficiently small. Then, (3.1)–(3.4) has a unique global smooth solution $(\rho, v, T, \phi)(x, t)$ *satisfying*

$$\|(\rho - 1, \rho_x, \rho_t, v, v_x, v_t, T - \overline{T}, T_x, T_t)\|_{C[0,1]}(t) \to 0 \quad \text{when } t \to \infty.$$

Proof We now use the continuation argument to prove the global existence. From the local existence and uniqueness theorem, there exists an interval $[0, t_0)$ such that (3.1)–(3.4) has a unique smooth solution. Then, Theorem 3.3 indicates that

$$E(t_0) \le \varepsilon/\Gamma.$$

At this point, we can apply the local existence and uniqueness theorem in previous sections to continue the solution to an interval $[t_0, t_1]$, $t_0 < t_1$. We can continue this process onto $[0, \infty)$ unless the sequence $\{t_n\}$ converges to a finite number $t^* < \infty$ as $n \to \infty$. For such a t^* we must have a maximal interval of existence $[0, t^*)$; i.e., (3.19) holds. This contradicts a priori estimate (3.35). Thus there can be no such $t^* < \infty$ and

$$\|(w_y, \theta)\|_{C([0,1] \times [0, \infty))} \le \varepsilon$$

holds. Then we have shown the existence of unique global smooth solution satisfying

$$(D^3 w, D^2 \theta) \in L^2([0, \infty); L^2(0, 1));$$
$$(D^2 w, D\theta)(y, t) \to 0 \quad \text{uniformly as } t \to \infty.$$

By (3.11), we obtain

$$\|(\rho - 1, \rho_x, \rho_t, v, v_x, v_t, T - \overline{T}, T_x, T_t)\|_{C[0,1]}(t) \to 0 \text{ when } t \to \infty.$$

This completes the proof of Theorem 3.4. □

4 Singular Relaxation Limit to Drift-Diffusion Equations

In this section we study the singular limit of (3.1)–(3.4) when the relaxation times τ_w and τ_p tend to zero with $0 \le 2\tau_w - \tau_p \le M\sqrt{\tau_p \tau_w}$. We first scale the variables (ρ, v, T, ϕ), and then show that the limit functions satisfy the drift-diffusion system as $\tau_p, \tau_w \to 0$.

Let

$$\rho^\tau(x,s) = \rho\left(x, \frac{s}{\tau_p}\right), \quad v^\tau(x,s) = \frac{1}{\tau_p} v\left(x, \frac{s}{\tau_p}\right),$$

$$\theta^\tau(x,s) = \frac{1}{\tau_w}\left(T\left(x, \frac{s}{\tau_w}\right) - \overline{T}\right),$$

$$\phi^\tau(x,s) = \phi\left(x, \frac{s}{\tau_p}\right), \quad d^\tau(x,s) = d\left(x, \frac{s}{\tau_p}\right).$$

Then (3.1)–(3.4) is transformed as follows:

$$\rho^\tau_s + (\rho^\tau v^\tau)_x = 0, \tag{4.1}$$

$$\tau_p^2 v^\tau_s + \tau_p^2 v^\tau v^\tau_x + \frac{1}{\rho^\tau}((\tau_w \theta^\tau + \overline{T})\rho^\tau)_x + v^\tau = \phi^\tau_x, \tag{4.2}$$

$$\tau_w^2 \theta^\tau_s - \frac{2\kappa\tau_w}{3\rho^\tau}\theta^\tau_{xx} + \tau_p\tau_w v^\tau \theta^\tau_x + \frac{2\tau_p}{3}(\tau_w\theta^\tau + \overline{T})v^\tau_x - \frac{\tau_p(2\tau_w - \tau_p)}{3\tau_w}(v^\tau)^2 + \theta^\tau = 0, \tag{4.3}$$

$$\phi^\tau_{xx} = \rho^\tau - d^\tau, \tag{4.4}$$

with the following initial-boundary data:

$$(\rho^\tau, v^\tau, \theta^\tau) = (\rho_0(x), v_0(x), T_0(x) - \overline{T}),$$
$$(v^\tau, \theta^\tau_x, \phi^\tau)(i,s) = 0, \quad i = 0,1, \quad s \geq 0.$$

We assume that initial data $(\rho_0(x), v_0(x), T_0(x) - \overline{T})$ are independent of τ_p and τ_w.

Theorem 4.1 *Assume that the conditions of Theorem 3.2 hold. Let (ρ, v, T, ϕ) be the solution of (3.1)–(3.4) constructed in Theorem 3.4. Then there exists $N(x,s)$ such that*

$$(\rho^\tau, \phi^\tau_x) \to \left(N, \int_0^1 \int_\eta^x (N(\xi,s) - d(\xi))d\xi d\eta\right), \quad \text{a.e. as } \tau_p, \tau_w \to 0.$$

The limit function N satisfies the drift-diffusion equation,

$$N_s - \left(N \int_0^1 \int_\eta^x (N(\xi,s) - d(\xi))d\xi d\eta - \overline{T} N_x\right)_x = 0,$$

in the sense of distributions.

Proof To show the convergence of subsequence $\{\rho^\tau, v^\tau, \theta^\tau\}$, we need some estimates which are independent of τ_p and τ_w.

Notice that the scaling sequence $(\rho^\tau, v^\tau, \theta^\tau, \phi^\tau)$ satisfies (4.1)–(4.4). Then the relations of the scaling sequence and the function (w, θ), satisfying (3.26)–(3.27), are as follows, where we use τ for both τ_p and τ_w:

$$\rho^\tau(x,s) = \frac{1}{w_y(y(x, \frac{s}{\tau}), \frac{s}{\tau}) + 1}, \qquad v^\tau(x,s) = \frac{1}{\tau} w_t\left(y\left(x, \frac{s}{\tau}\right), \frac{s}{\tau}\right),$$

$$\theta^\tau(x,s) = \frac{1}{\tau}\theta\left(y\left(x, \frac{s}{\tau}\right), \frac{s}{\tau}\right),$$

$$\rho^\tau_x(x,s) = -(\rho^\tau)^3 w_{yy}\left(y\left(x, \frac{s}{\tau}\right), \frac{s}{\tau}\right), \qquad v^\tau_x(x,s) = \frac{1}{\tau}\rho^\tau w_{yt}\left(y\left(x, \frac{s}{\tau}\right), \frac{s}{\tau}\right),$$

$$\rho^\tau_{xx}(x,s) = -(\rho^\tau)^4 w_{yyy}\left(y\left(x, \frac{s}{\tau}\right), \frac{s}{\tau}\right) + 3(\rho^\tau)^5 \left(w_{yy}\left(y\left(x, \frac{s}{\tau}\right), \frac{s}{\tau}\right)\right)^2,$$

$$\phi^\tau(x,s)_x = \int_0^1 (w_y + 1) \left(\int_\eta^{y(x, \frac{s}{\tau})} \left(1 - (w_y + 1)\tilde{d}\right)\left(\xi, \frac{s}{\tau}\right) d\xi\right) d\eta,$$

$$\phi^\tau(x,s)_{xx} = \rho^\tau(x,s) - d(x),$$

$$\phi^\tau(x,s)_{xs} = -(\rho^\tau - d)v^\tau(x,s) + \int_0^1 \frac{w_{yt}}{\tau} \left(\int_\eta^y \left(1 - (w_y + 1)\tilde{d}\right)\left(\eta, \frac{s}{\tau}\right) d\eta\right) dy$$

$$- \int_0^1 (w_y + 1) \left(\int_\eta^y \frac{1}{\tau}\left(w_{yt}\tilde{d} - \tilde{d}_x(w_y + 1)w_t\right)\left(\xi, \frac{s}{\tau}\right) d\xi\right) d\eta.$$

Therefore,

$$\|(\rho^\tau, \phi^\tau, \phi^\tau_x)\|_{L^\infty} + \int_0^\infty\int_0^1 [(v^\tau)^2 + (\theta^\tau)^2 + (\rho^\tau_s)^2 + (\rho^\tau_x)^2 + (\phi^\tau_{xx})^2 + (\phi^\tau_{xs})^2] dy\,ds \le C,$$

where $C > 0$ is independent of τ.

Thus, there exists $N(x, s)$ such that

$$(\rho^\tau, \phi^\tau_x) \to (N, \Phi_x) \quad \text{a.e. as } \tau \to 0,$$

where $\Phi_x(x,s) = \int_0^1 \int_\eta^x (N(\xi, s) - d(\xi)) d\xi\,d\eta$.

From (3.1)–(3.2),

$$\int\int (\rho^\tau \psi_s + \rho^\tau v^\tau \psi_x) dx\,ds = 0,$$

$$\tau_p \int\int (\rho^\tau v^\tau \psi_{xs} + \tau_p \rho^\tau (v^\tau)^2 \psi_{xx}) dx\,ds + \int\int \rho^\tau (\tau_w \theta^\tau + \overline{T}) \psi_{xx} dx\,ds$$

$$= \int\int (\rho^\tau v^\tau - \rho^\tau \phi^\tau_x) \psi_x dx\,ds,$$

for all $\psi \in C_0^\infty(R \times R_+)$. That is, for any test function $\psi(x,t) \in C_0^\infty((0, 1) \times (0, \infty))$, the relation,

$$\int\int (\rho^\tau \psi_s + \rho^\tau \phi^\tau_x \psi_x + \overline{T} \rho^\tau \psi_{xx}) dx\,ds$$

$$= -\int\int (\tau_p \rho^\tau v^\tau \psi_{xs} + (\tau_p^2 \rho^\tau (v^\tau)^2 + \tau_w \rho^\tau \theta^\tau)\psi_{xx}) dx\,ds,$$

holds. Let $\tau_p, \tau_w \to 0$. Since

$$\|\rho^\tau\|_{L^\infty((0,1)\times(0,\infty))} + \|(v^\tau,\theta^\tau)\|_{L^2((0,1)\times(0,\infty))} \leq C,$$

where C is independent of τ_p, τ_w, and t, then

$$\int\int (N\psi_s + N\Phi_x\psi_x + \overline{T}N\psi_{xx})dxds = 0.$$

This completes the proof of Theorem 4.1. □

Acknowledgements

The first author is supported by the Office of Naval Research under grant N00014-91-J-1384, the National Science Foundation under grant DMS-9623203, and by an Alfred P. Sloan Foundation Fellowship. The second author is supported by the National Science Foundation under grant DMS-9424464.

Bibliography

1. Baccarani, G. and Wordeman, M. R. (1985). An investigation of steady–state velocity overshoot effects in Si and $GaAs$ devices. *Solid State Electronics*, **28**, 404–416.
2. Barcilon, V., Chen, D., Eisenberg, R., and Jerome, J. (1997). Qualitative properties of solutions of steady-state Poisson–Nernst–Planck systems: perturbation and simulation study, *SIAM J. Appl. Math.*, **57**, 631–648.
3. Bløtekjær, K. (1970). Transport equations for electrons in two–valley semiconductors. *IEEE Trans. Electron Devices*, **17**, 38–47.
4. Cercignani, C. (1987). *The Boltzmann Equation and its Application*, Springer.
5. Chen, D., Eisenberg, R., Jerome, J., and Shu, C.-W. (1995). Hydrodynamic model of temperature change in open ionic channels. *Biophysical Journal*, **69**, 2304–2322.
6. Chen, G.-Q. (1992). Global solutions to the compressible Navier–Stokes equations for a reacting mixture. *SIAM J. Math. Anal.*, **23**, 609–634.
7. Chen, G.-Q., Jerome, J. W., Shu, C.-W. and Wang, D. (1996). Two-carrier semiconductor device models with geometric structure and symmetry properties. *Chapter 6 in the present volume*.
8. Chen, G.-Q., Jerome, J. W., and Zhang, B. (1997). Particle hydrodynamic moment models in biology and microelectronics: singular relaxation limits. *Nonlinear Anal.*, **30**, 233–244.
9. Dafermos, C. M. and Nohel, J. A. (1979). Energy methods for nonlinear hyperbolic Volterra integrodifferential equations. *Comm. PDE*, **4**, 219–278.

10. Degond, P., Guyot-Delaurens, F., Mustieles, F.J., and Nier, F. (1990). Semiconductor modelling via the Boltzmann equation. In: *Mathematical Modelling and Simulation of Electrical Circuits and Semiconductor Devices* (eds. R. E. Bank, R. Bulirsch, and K. Merten), pp. 153–167. Birkhäuser.
11. Fatemi, E., Jerome, J. W., and Osher, S. (1991). Solution of the hydrodynamic device model using high–order non–oscillatory shock capturing algorithms. *IEEE Trans. on Computer-Aided Design*, **10**, 232–244.
12. Gardner, C. L. (1991). Numerical simulation of a steady-state electron shock wave in a submicron semiconductor device. *IEEE Trans. on Electron Devices*, **38**, 392–398.
13. Gardner, C. L., Jerome, J. W., and Rose, D. J. (1989). Numerical methods for the hydrodynamic model: subsonic flow. *IEEE Trans. on Computer-Aided Design*, **8**, 501–507.
14. Jerome, J. W. (1996). *Analysis of Charge Transport: A Mathematical Study of Semiconductor Devices*. Springer.
15. Jerome, J. W. and Shu, C.-W. (1995). Transport effects & characteristic modes in the modeling & simulation of submicron devices. *IEEE Trans. on Computer-Aided Design*, **14**, 917–923.
16. Jerome, J. W. and Shu, C.-W. (1995). The response of the hydrodynamic model to heat conduction, mobility, and relaxation expressions. *VLSI Design*, **3**, 131–143.
17. Klainerman, S. (1980). Global existence for nonlinear wave equations. *Comm. Pure Appl. Math.*, **33**, 43-101.
18. Lions, J. L. and Magenes, E. (1972). *Non-homogeneous Boundary Value Problems and Applications*. Springer.
19. Majda, A. (1984). *Compressible Fluid Flow and Systems of Conservation Laws in Several Space Variables*. Springer.
20. Marcati, P. and Natalini, R. (1995). Weak solutions to a hydrodynamic model for semiconductors & relaxation to the drift-diffusion equation, *Arch. Rat. Mech. Anal.*, **129**, 129–145.
21. Mastumura, A. and Nishida, T. (1980). The initial value problem for the equations of motion of viscous and heat-conductive gases. *J. Math. Kyoto Univ.*, **20**, 67–104.
22. Park, J. and Jerome, J. (1997). Qualitative properties of solutions of steady-state Poisson–Nernst–Planck systems: mathematical study, *SIAM J. Appl. Math.*, **57**, 609–630.
23. Patton, H., Fuchs, A, Hille, B., Scher, A., and Steiner, R. (eds) (1989). *Textbook of Physiology*, 21st edn, vol. 1, pp. 1–47. W.B. Saunders.
24. Rubinstein, I. (1990). *Electro-Diffusion of Ions*, SIAM Studies in Applied Mathematics. SIAM.
25. Rudan, M., Odeh, F., and White, J. (1987). Numerical solution of the hy-

drodynamic model for a one-dimensional semiconductor, *COMPEL (The International Journal for Computation and Mathematics in Electronic Engineering)*, **6**, 151–170.
26. Selberherr, S. (1984). *Analysis and Simulation of Semiconductor Devices*. Springer.
27. Slemrod M. (1981). Global existence, uniqueness and asymptotic stability of classical smooth solutions in one-dimensional, nonlinear thermoelasticity. *Arch. Rat. Mech. Anal.*, **76**, 97–133.
28. Van Roosbroeck, W. (1950). Theory of flow of electrons and holes in germanium and other semiconductors. *Bell System Tech. J.*, **29**, 560–607.
29. Wang, D. and Chen, G.-Q. (1996). Formation of singularities in compressible Euler-Poisson fluids with heat diffusion and damping relaxation. *J. Differential Eqs.*, to appear.
30. Zhang B. (1995). Global existence and asymptotic stability to the full 1D hydrodynamic model for semiconductor devices. *Indiana Univ. Math. J.*, **44**, 971–1005.